# North Carolina Fire Law

# North Carolina Fire Law

THIRD EDITION

C. Barrett Graham

Carolina Academic Press
Durham, North Carolina

Library of Congress
Cataloging-in-Publication Data

Names: Graham, C. Barrett, author.
Title: North Carolina fire law / by C. Barrett Graham.
Description: Third edition. | Durham, North Carolina: Carolina Academic Press, LLC, [2019] | Includes bibliographical references and index.
Identifiers: LCCN 2019031999 | ISBN 9781531017385 (paperback) | ISBN 9781531017392 (ebook)
Subjects: LCSH: Fire departments—Law and legislation—North Carolina. | Fire fighters—Legal status, laws, etc.—North Carolina. | Fire prevention—Law and legislation—North Carolina.
Classification: LCC KFN7781.G73 2019 | DDC 344.75605/377—dc23
LC record available at https://lccn.loc.gov/2019031999

CAROLINA ACADEMIC PRESS
700 Kent Street
Durham, North Carolina 27701
Telephone (919) 489-7486
Fax (919) 493-5668
www.cap-press.com

Printed in the United States of America

Dedicated to:

Sharyne,
*my wife;*

Ruby B. Graham
(1916–2012)
and
John B. Graham, MD
Alumni Distinguished Professor of Pathology
Professor Emeritus
University of North Carolina-Chapel Hill
(1918–2004),
*my parents*;

and
North Carolina's Firefighters

# Contents

# Preface

To the student:

Please keep the following in mind when working with these materials.

1. This book is not intended as a detailed discussion of the law applicable to the fire service (lawyers call such books "hornbooks"); rather, it is an orientation and survey of guideposts for reference and further study. It consists of an overview of North Carolina and federal law.

2. Each state has its own laws and regulations governing its fire service-related activities. This means that when you cross a state line, the rules will change. The law discussed herein, for the most part, is that which is applicable only in North Carolina.

3. Legislative activities and court rulings have a continuous influence on law. Consequently, the law is subject to frequent, and often unannounced, changes. As a result, a student of law must not assume that what the law is today will remain the law forever. A serious student of a body of law must monitor it constantly and identify the changes in order to use it effectively, or, at the very least, the student must check for changes before making an important decision based upon what the law or rule "used to be."

4. This text is intended to provide effective research tools for both the first-time student and the supervisor.

# Acknowledgments

This project represents the most recent step in an academic adventure which began in 1991 when the late Emmel Coggins (former firefighter, Fire Service Instructor, regional training coordinator, and fierce advocate for North Carolina's Community Colleges) shanghaied me into writing a course for firefighters dealing with civil liability. Emmel, you may have created a monster. Thanks, old man, wherever you are, for your advice and support. Thanks are due to my wife, Sharyne, for her patience, support, and occasional boots to the rear, all of which combined to keep this project moving; to two distinguished professors, the late John B. Graham, MD, and the late John G. Barrett, PhD, who unhesitatingly encouraged me to pursue this project at its beginning; and to my parents, who saw to it that I received a decent education, my efforts to the contrary notwithstanding.

In my wanderings through academia, one person who was to have a profound effect upon on many of us who studied under him was the late Franklin "Frank" Butler. He taught demanding courses in English, writing, and literature at Asheville School (where I graduated on schedule after four years and two summer schools). That which I have done correctly in this book safely can be attributed to Frank's influence. The errors are my fault entirely.

Finally, thanks to the folks at Carolina Academic Press for their assistance in this project and their patience with a rookie author, and to Beverley Guldner and Marilyn Gould for word-processing assistance during the writing and editing of the manuscript of the first edition of this book.

# North Carolina Fire Law

## Chapter 1

# Organization and Sources of the Law

To understand any legal system a student should have an overview of its physical structure and organization. This chapter will give the student an overview of the federal and North Carolina systems.

An examination of the system begins with the Constitution of the United States. All state and federal laws, state constitutions, statutes, rules, regulations, and ordinances must follow certain basic principles in the Constitution of the United States in order to be enforceable. This means that the original source of all law for all citizens of the United States and persons or entities under the jurisdiction of the courts of the United States, the states, the territories, and any of their respective political subdivisions is the Constitution of the United States. Thus, the conduct of citizens and residents of North Carolina is governed by two bodies of law—state law and federal Law.

## Federal Law

Federal law is based upon the Constitution, which consists of two parts. The first part is composed of the first seven articles. These articles establish and empower the federal government as we know it today. The second part of the Constitution consists of the amendments, of which there are twenty-seven. The first ten amendments are known popularly as the Bill of Rights and were created by the men who wrote the Constitution when they recognized the need to describe the rights and freedoms which the federal government was created to protect. One additional amendment related to the Bill of Rights was ratified by the states in 1868, following the War Between the States, in order to protect the rights of citizens at both state and federal levels of governance. The Fourteenth Amendment confirms that the Constitution applies to all of us all of the time, at all levels of government.

The first seven articles provide for three branches of government—the executive, the legislative, and the judicial. The Constitution is intended to keep these three basic branches of government in existence at all levels of government throughout the United States. The men who wrote the Constitution wanted to avoid a centralized form of government and, accordingly, divided the powers of government between the three branches, including specific language describing the branches' respective duties. This concept is known as the "separation of powers" and has been carefully protected by the courts throughout the history of the United States.

The executive branch of government enforces or executes the law. At the federal level, the executive branch consists of the president and the cabinet. The cabinet consists of the leaders of the various agencies of the federal government—for example, the Secretary of Defense, the Attorney General, and the Secretary of Commerce. The legislative branch of government creates the law. At the federal level, the legislative branch consists of the Senate and the House of Representatives, collectively referred to as Congress. The judicial branch of government interprets the law. At the federal level, the judicial branch consists of the federal court system, with the Supreme Court of the United States at the top of the system.

A unique concept integrated into the Constitution by its framers as a control over the three branches of the federal government is called "checks and balances." In an effort to prevent any one of the three branches from dominating either of the other two, the Constitution gives each branch of government some control over the other two, either directly or indirectly. For example, while a federal judge is able to declare laws unconstitutional, and therefore, unenforceable, that same judge must be appointed by the president and confirmed by the Senate, and Congress can remove the judge from office because of misconduct.

Similarly, while the president is the senior law enforcement official in the United States and the commander in chief of the armed forces, Congress can remove the president from office by impeachment by the House of Representatives (indictment) and a trial before the Senate, with the presiding judge being the Chief Justice of the Supreme Court. The most recent example of this was the impeachment and trial of Bill Clinton. Members of Congress are subject to the authority of the Constitution, as well. They may be removed by their peers; the laws they enact may be declared unconstitutional by the courts and therefore void or unenforceable; and as citizens, they remain subject to the authority of the executive branch of government and the judicial system.

The body of federal law next below the Constitution is the statutory law enacted by Congress. This body of law is known as the United States Code. These laws are organized into chapters called "Titles." Each Title, is, in turn, divided into individual statutes. In order to find a particular statute in the United States Code, the process is relatively simple if the searcher knows the applicable numbers. They are organized with a first number, the Title, followed by U.S.C for "United States Code," and a number, representing the statute itself. The title in the United States Code which contains all the basic laws regarding bankruptcy is Title Eleven, for example, and a portion of the Bankruptcy Code would be identified as 11 U.S.C. 362. A large portion of the civil rights law enacted by Congress is found in Title 42 of the United States Code. One of the most frequently referenced civil rights statutes is 42 U.S.C. 1983, and it is often used as one of the legal bases in a lawsuit involving civil rights.

Once a statute has been enacted by Congress, it becomes necessary, upon occasion, to enact rules and regulations to explain how the particular statute is supposed to be implemented. These are contained in the Code of Federal Regulations (CFR). The CFR is organized into titles and chapters with numbers and names, not unlike the United States Code. They are published, ordinarily, by the agency of the executive branch of government which has been given the responsibility for the enforcement of the statutory law—for example, the Environmental Protection Agency, the Department of Veterans Affairs, the Department of Labor, and so forth. In the fire service, the two bodies of rules and regulations we are encountering most frequently are Hazardous Materials (HAZMAT) and the Fair Labor Standards Act (FLSA). When a dispute involves one or more of the rules in the CFR, the matter is referred to one or more types of hearing boards or commissions, whose job is to review the law and facts regarding the incident(s) and make one or more rulings. Should either side in one of these disputes decide to appeal a decision, there exists a series of hearings and appeals within the CFR which must be followed before the case can go to court. These rules and regulations and the legal work performed to implement and interpret them are generally referred to as administrative law.

When attempts to enforce the CFR began in earnest, those who opposed the provisions of the CFR went to court to stop the process. The argument was made that the CFR was unconstitutional, for the most part, because the regulations had not been enacted by Congress. They had been enacted by the executive branch of government, which, under the Constitution, has authority only to enforce the law. A strong argument was made that the creation of the CFR had violated the separation-of-powers concept in the Constitution. The courts ruled, however, that "practical necessity" required that the agencies of the executive branch of government issue the regulations and have allowed the practice to continue to this day. However, the courts required that once issued, the regulations would have to be voted upon and approved (that is, ratified) by Congress before they could be enforced. Ratification would thus convert the regulations into enforceable laws. The courts pointed out that it was more efficient for the agency with enforcement responsibility to write the rules because the agency possessed the expertise. Additionally, the court system would be available to issue rulings on regulations which might violate basic constitutional principles. Some recent activities by the courts relating to the fire service and the FLSA are excellent examples of this process in action.

Article III of the Constitution places the judicial power of the United States in the Supreme Court, as well as in such other inferior courts as "Congress may ordain and establish." The role of the courts in relation to the laws of the United States was not spelled out in the Constitution, so it became one of the tasks of the early Supreme Court to determine its role as the third branch of the federal government. Legal scholars recognize one chief justice of the Supreme Court during its formative years as the most important—John Marshall. In 1803, his court handed down a decision that most scholars regard as possibly the most important constitutional law case in American history. The case, *Marbury v. Madison*, was a dispute concerning whether or not a federal court could order a government official to comply with the law. The dispute itself was not nearly as important as Chief Justice Marshall's ruling in the case. With his ruling, Chief Justice Marshall established the concept of *judicial review* in the court system of the United States. Judicial review is the concept of comparing a law against the Constitution to determine if the law in question meets certain constitutional requirements. If it does not, it is declared unconstitutional and cannot thereafter be enforced. Because the Constitution, in Article VI, makes the Constitution and federal laws the supreme law of the land and, in Article III, gives the federal courts jurisdiction over all cases involving the supreme law of the land, Chief Justice Marshall's ruling in *Marbury v. Madison* gave the federal judiciary its role in government—to review the law, interpret it, and then to issue rulings based upon the review and interpretation. This concept was adopted by the states as they formed their individual court systems and remains alive and well today. Judicial review gives the judicial branch of government the authority it needs to balance it with the executive and legislative branches. Today, the interpretation of a law by a court is regarded as the ultimate meaning of the law, regardless of what a legislative body may have intended by its language of that law.

In 1819, Marshall issued another Supreme Court opinion that helped set in place a second basic rule by which the courts in the United States operate today. In *McCullough v. Maryland*, a dispute over the incorporation of a bank in Maryland, the Supreme Court ruled that the Constitution and the laws enacted by Congress in accordance with it were to be the supreme law of the land and could not be modified by the states. The federal judiciary then became the highest-ranked court system in the land, and the Constitution and federal law became the supreme law of the land.

## North Carolina Law

North Carolina law originates with the Constitution of North Carolina. The provisions of the North Carolina constitution remain subject to those principles set out in the Constitution of the United States, so the supreme law concerning a citizen and/or resident of North Carolina is the U.S. Constitution unless the legal question in dispute is one addressed solely by the Constitution of North Carolina. If the question is solely a North Carolina one, then the final constitutional answer will be the North Carolina one unless there is a conflict between the Constitution of North Carolina and the Constitution of the United States. The current version of the North Carolina constitution was ratified by referendum in 1970 and became effective July 1, 1971. It is a substantial revision of its predecessor, the North Carolina Constitution of 1868.

The North Carolina constitution consists of fourteen articles. Article I is entitled "Declaration of Rights" and contains a list of thirty-seven rights which are guaranteed to citizens and/or residents of North Carolina. Many of these rights are found in the first ten amendments to the Constitution of the United States and several of the others clearly are the result of the War Between the States. Article II establishes the state's legislative power, which is vested in the General Assembly, which consists of the House of Representatives and the Senate; and as in the Constitution of the United States, Article II deals with various other laws concerning the establishment and management of the General Assembly. Article III places the executive power of North Carolina with the governor and proceeds thereafter to describe the duties, powers, and responsibilities of the office. Article IV places the judicial power of the state in the North Carolina court system. Thus, in much the same fashion as the Constitution of the United States, the constitution of North Carolina has established three branches of government—legislative, executive, and judicial—with each branch serving essentially the same function as its federal counterpart.

Amendments to the constitution of North Carolina begin in the General Assembly and, once ratified by the General Assembly, must be offered to the voters of North Carolina for approval. When a majority of North Carolina's voters have voted in favor of the amendment, it becomes law and is incorporated into the North Carolina constitution. All other published statewide law in North Carolina must originate with the General Assembly as well. The primary body of published law in North Carolina is known as the North Carolina General Statutes. Three fairly common abbreviations are used as prefixes for the General Statutes—NCGS, GS, and GSNC. They apply to all citizens, residents, and other persons or entities who are subject to the jurisdiction of the North Carolina courts.

The General Statutes are found in book form in libraries which may have a need for them and can be accessed on the Internet through appropriate Web sites made available through agencies of the North Carolina state government. The books are published in green covers and may be loose-leaf or fully bound. A complete set of the General Statutes is *very expensive* and requires expensive annual updating in order to be kept current. Copies of the General Statutes can be found in law offices, courthouses, public libraries, law school libraries, and community college libraries. The latter may, in fact, be the best place to locate updated copies of the General Statutes. They can be found in the reference section. The General Statutes are arranged numerically and by topic, but not alphabetically, and, therefore, you must know either the topic you wish to research or the number of the statute in order to locate the appropriate volume and statute.

Presently there are two numbering systems for the General Statutes. The first is the older, two-number system. An example is GS 20-140. The first number indicates the chapter in the General Statutes where the particular statute is found, and the second number indicates the statute itself. Both numbers are necessary for identifying the statute correctly. In the example above, Chapter 20 is the chapter of the General Statutes which governs motor vehicles. The number "140" is the statute in Chapter 20 which covers reckless driving. The second numbering system has three numbers. An example is GS 58-82-1. This systems works the same as the other one. The only difference is that an article (subchapter) number is inserted into the sequence in order to make available a larger range of numbers for future use. Chapter 58 contains most of the statutes relating to insurance matters. Article 82 (which is in reality a subchapter) contains some of the law relating to the fire service. The third number is the statute itself. This particular statute gives the fire service the authority to act on the fire ground and provides a mechanism for dealing with those who interfere with an ongoing fire ground evolution.

The spine of each book displays the numerical range of the statutes contained in the book. So, if you know the number, you need only examine numbers on the spines of the books until you find the correct book. A search becomes more complex when you have a topic but no number. There is an alphabetical index to the General Statutes, located in two or more volumes after the last volume of the statutes. Unfortunately, this index is difficult to use and appears to have been prepared by someone who does not have a good understanding of the General Statutes. However, each chapter in the General Statutes begins with a fairly comprehensive table of contents. So sometimes it is possible to succeed in a search by examining the table of contents and then examining likely statutes based upon it. If this procedure does not succeed, try contacting either the lawyer representing your organization or the Institute of Government (School of Government) at the University of North Carolina at Chapel Hill. The School of Government has staff members who specialize in most areas of governmental law, including the fire service, and these specialists can be of considerable assistance.

As a general rule, the General Statutes control the conduct of individuals and other entities, telling them what they can or cannot do. Sometimes, however, the statute does not carry sufficient instructions regarding a particular activity. If instructions are insufficient, often there exists a regulation regarding the statute which gives more guidance about the activity. These regulations are North Carolina's equivalent of the CFR and are found in the North Carolina Administrative Code. Ordinarily these regulations, or administrative laws, are written by the agency of state government which oversees enforcement of laws regarding the activity in question. These regulations carry the full legal force of the General Statutes once they have been ratified by the legislature. The North Carolina Administrative Code is ordinarily abbreviated as NCAC and is divided into mega-chapters, called "Titles" and subdivided into chapters, and sections. When researching a portion of the NCAC, begin with a title and then move to a chapter and, finally, to a section. For example, the portion of the NCAC covering basic rules for the North Carolina Code Officials Qualification Board (the licensing board for all building code inspectors, including fire inspectors) is found in Title 11, NCAC, Chapter 8, Sections .0500 through .0836.

Sometimes, a municipal or county government will have an issue which is peculiar to that locality and no other. Because the issue is not statewide, a General Statute may not be an appropriate mechanism to deal with the issue. However, the legislature can still deal with it by utilizing what is known as a "local act." Because it is passed by the legislature, a local act has the force of a General Statute, but it deals only with the particular issue within the jurisdiction of the municipality or county named in the act. Hunting regulations and local modifications to the building code are common examples. The author was involved directly in a local act some years ago.

In the early 1990s a problem arose regarding the disposition of some funds which had been donated to the author's fire department. The department was a municipal one staffed entirely by volunteers and covering a sizeable rural fire district as well as the municipality. The donated funds were to become part of the general fund of the town, but the volunteers and certain members of the public wanted the funds reserved for fire protection only. The then-sitting members of the town board of commissioners were able to guarantee the use of the funds for as long as they were in office, but no longer, so the parties involved sought a mechanism to assure the reservation of the funds for fire protection should there be any funds remaining after any changes in the membership of the board of commissioners. The result was a local act, passed by the General Assembly, which established a mechanism for managing the funds. Because the local act was passed, it would be necessary to return to the General Assembly in order to expend the funds in any manner other than what was specified in the original local act. This solution to the original problem worked, and the funds were eventually expended entirely for fire protection.

The final (lowest) level of legislation in North Carolina consists of municipal and county ordinances. These are laws which are passed by the counties and municipalities themselves covering local issues. Local ordinances cannot be used to modify General Statutes or administrative codes unless the modifications are approved by either the General Assembly or the applicable administrative agency, as the case may be. Ordinances cover a wide variety of situations, including budgets, parking, code enforcement, housing conditions, speed limits, and many other activities which affect the lives of the inhabitants of the counties and municipalities.

In the management of the law, the role of the North Carolina court system, known officially as the General Court of Justice, is extensive and far-reaching and is based upon the authority of the courts to interpret the law, their primary constitutional responsibility. This means that the courts must tell the citizens of North Carolina what the law means. The legislature may have enacted a law, but the courts' interpretations of that law become its ultimate meaning. These interpretations are the last vestiges of the English common law, which is the philosophical basis for most North Carolina law.

In simplified form, the common law was unwritten law carried in the minds of presiding judicial officials which was enforced as though it were written. The basic idea was that if a certain type of behavior were to be followed and there was no objection to it, then it became lawful. Eventually, English authorities on the common law began to write commentaries on it, and these commentaries, together with much of the unwritten common law, came to the colonies with the English settlers. The desire of the colonists to put the common law into writing and adapt it to their needs led eventually to what we now know as the General Statutes. The commentaries by the English writers were replaced by the written opinions of the courts. Small bits of the unwritten common law are still followed in North Carolina and are found mostly in real property (land) law and contract law.

Occasionally, a court may issue a ruling with such a sweeping effect that the court is accused of creating an original law. Because the creating of law is the responsibility of the legislative branch of government, such rulings are often appealed because the party appealing believes that the ruling of the court violates the requirements of separation of powers specified in the constitutions of the United States and North Carolina. Sometimes the problems created by these rulings are dealt with by a higher court; sometimes they are dealt with by the legislature; and sometimes they are left in place until a similar case is presented to the court in question and it can reconsider its earlier decision.

## Chapter 2

# The Courts System

This chapter is an overview of the court systems with jurisdiction over North Carolina. *Jurisdiction* means the geographical area or subject matter over which a government or one of its agencies, for example, a court system, is able to exercise its authority. It is vital for a lawyer to examine and understand the operation of a court system—especially for a lawyer planning a certain type of legal action. One court system may appear to be more appropriate than another because of its efficiency, subject-matter jurisdiction, its ability to issue certain types of rulings, or some other factor. Sometimes, two or more court systems exercise jurisdiction over a particular subject matter. This situation is known as *concurrent jurisdiction*. When a matter is restricted to the jurisdiction of a single court or court system, that situation is known as *exclusive jurisdiction.* For example, both North Carolina and Federal courts have jurisdiction over civil rights cases (concurrent jurisdiction), but only the United States Bankruptcy Court can allow someone to file bankruptcy (exclusive jurisdiction).

An important decision by a court is issued in writing and is known as an *opinion* or *ruling* (*opinion* is the most commonly used term). A written opinion is the court's statement of what the law means and is frequently utilized by lawyers and courts in similar cases. When a court issues an opinion in a case, we see the application of the same principles which led to the English common law centuries ago. The primary difference is that in American courts the opinion is issued in writing and thus made available for study and future use.

## Federal Courts

The federal court system is the senior court system for citizens and residents of the United States and its territories. It consists of two basic divisions: a trial division, which manages cases through their various stages of trials, and an appellate division, which reviews cases during and after trial to determine if proper law and procedure were followed at trial. The trial division consists of the United States District Courts and a number of specialized courts with limited subject-matter jurisdiction, such as tax issues, claims against the United States admiralty, and bankruptcy.

The primary trial courts in the federal system are the United States district courts. For administrative purposes, they are organized into groups of states called Circuits, which are identified by numbers. North Carolina is part of circuit number four (Fourth Circuit). It consists of Maryland, Virginia, West Virginia, North Carolina, and South Carolina. Each circuit has a court of appeals, which reviews cases within its respective circuit and is known as the United States Court of Appeals for that circuit. The United States Court of Appeals for the Fourth Circuit serves North Carolina. It is headquartered in Richmond, Virginia, and consists of fifteen judges who hear and review cases in groups of three judges except in unusual circumstances, when a larger group may review a case. Below the circuit courts of appeals are the United States District Courts. The judges who preside in these courts are known as United States District Judges, and they have wide-ranging authority. The district courts are organized by individual states, and, if the state has sufficient population, the district courts are subdivided into districts (groups of counties) within the state. In North Carolina, they are the United States District Courts for the Eastern, Middle, and Western Districts of North Carolina.

Also within the United States District Dourts are United States magistrate judges. These judges are subordinate to the United States district judges and help manage the court caseload by handling some of the district court cases before they reach the district court judges. Another group of judges who work within the federal court system in North Carolina are the United States Bankruptcy Judges. Their courts are organized geographically in North Carolina in exactly the same way as the United States District Courts, with bankruptcy courts in each district of the state. The bankruptcy judges are specialized, with their authority, which is great, limited to matters involving the Bankruptcy Code.

The court which is the ultimate authority for any case arising in North Carolina is the United States Supreme Court. Composed of nine justices, it is headquartered in Washington, D.C., and is the final level of appeal for all legal cases initiated within the United States court system or the court system of any state. Thousands of cases are submitted to the Court each year (in excess of 50,000, according to the Office of the Clerk), and of this number, barely 125 are selected by the Court for consideration.

In accordance with the Constitution of the United States, all judges and justices in the federal court system are nominated by the president, and their nominations are approved by the Senate. This process involves exhaustive screenings of the candidates, resulting in the high quality of the eventual appointee. Also in accordance with the Constitution, most judges and justices in the federal courts system are appointed for as long as their behavior is good (for life), are paid handsome salaries, and have their expenses, courtrooms, offices, and staffs paid for by the taxpayer, all in an effort to avoid outside influences upon the decision-making process. One group of federal judges, called Title VIII judges because of the law which created the positions, are not appointed for life but for a term of years. Among this group are the bankruptcy and magistrate judges.

Each United States District Court and United States Bankruptcy Court has a headquarters managed by a clerk known in North Carolina as the Clerk of the United States District Court, or United States Bankruptcy Court, for the Eastern, Middle, or Western District of North Carolina. All cases must be filed initially with the appropriate clerk. After the filing of appropriate documents and the payment of required filing fees, the case is referred to the appropriate judge for further proceedings.

## North Carolina Courts

Between 1968 and 1970, the North Carolina court system was reorganized into the configuration we know today. As in the federal system, the North Carolina courts are organized into two divisions, a trial division and an appellate division. The trial division has three levels and the appellate division has two levels. An appellate division has statewide jurisdiction, while the trial division has its jurisdiction controlled at two levels geographically, with one level having statewide jurisdiction. The North Carolina court system is known as The General Court of Justice and has an administrative organization which manages it and provides logistical support, known as the Administrative Office of the Courts (AOC).

The Supreme Court of North Carolina is the senior appellate court in the state and the state's highest court. An appeal from the Supreme Court of North Carolina can be made only to the Supreme Court of the United States. The Supreme Court of North Carolina has seven justices who sit as a single group to hear cases and is headquartered in Raleigh. The junior appellate court in North Carolina is the North Carolina Court of Appeals. It consists of fifteen judges who sit in groups of three judges to hear the cases presented to them.

The senior trial division of the General Court of Justice is the Superior Court Division. It is composed of the superior court judges and the clerks of the superior court. The intermediate trial division of the North Carolina courts is the District Court Division and is composed of the district court judges. The junior trial division is the Magistrate or Small Claims Court. It corresponds roughly to the famous *Peoples Court* of television fame. The Small Claims Court is part of the district court system, and the presiding officers in Small Claims Court (magistrates) are appointed by the chief resident superior court judges with jurisdiction over the counties in which the magistrates will be holding court.

Superior court judges have jurisdiction over criminal felonies, criminal appeals from the District Court, civil cases where the dispute involves more than $25,000, and certain types of cases specified in the General Statutes. The superior court also contains a subdivision called the Business Court, which was created to deal with complex business-related lawsuits and which is staffed by selected superior court judges who have expertise dealing with such cases. The clerks of superior court are North Carolina's probate judges. They have the initial jurisdiction over the probate of wills, the administration of the estates of deceased persons, trusts,guardianships, incompetency proceedings, foreclosures, and several other categories of legal actions known as *Special Proceedings*. Appeals from rulings by the clerk of superior court are taken to superior court judges, unless a statute states otherwise.

District court judges have jurisdiction over criminal misdemeanors, civil cases involving $25,000 or less, juvenile criminal cases, family law matters, and other classifications of cases as specified in the General Statutes, including appeals from Small Claims Court. Appeals of criminal matters from District Court go to the Superior Court and appeals of civil matters go to the Court of Appeals.

Small claims (Magistrate) courts cover civil matters involving $10,000 or less, landlord/tenant matters, certain types of repossessions of property, and some other civil matters specified in the General Statutes. Magistrates, the presiding officers in small claims courts, also have limited jurisdiction in certain criminal matters. Among other duties, they can issue search warrants, set bail for arrested persons, and receive pleas of guilty in certain misdemeanor cases.

All judges and justices in North Carolina, with the exceptions of clerks of the superior court and magistrates, must be North Carolina lawyers in order to hold office. All judges and justices in North Carolina, with the exceptions of magistrates and certain district and superior court judges who are appointed by the governor, must be elected. Terms of office vary according to the office to which the candidate has been elected.

## The Role of the Courts

The basic role of the court systems is to interpret the law. Beyond that, the two divisions of the court system have slightly different roles. The trial division interprets the law and takes direct action based upon those interpretations (trials). The appellate division interprets the law and issues written opinions regarding the actions taken by the lower courts, sometimes agreeing, sometimes disagreeing, and sometimes partially agreeing or disagreeing. These interpretations are intended to be descriptions of what the law in a certain case is supposed to mean. Sometimes these rulings can have extraordinary consequences, as in the opinions of Justice John Marshall mentioned in Chapter One, and sometimes they can be informative and entertaining. An entertaining example is a 1998 case from the Supreme Court of North Carolina, *State v. Fly*. This case was an appeal from criminal court through the Court of Appeals to the Supreme Court involving a matter of indecent exposure. Mr. Fly had been accused of indecent exposure by displaying his posterior to a young lady in the stairwell of an apartment complex. After a discussion of the applicable statute, the facts of the case, and matters concerning swimsuit design, the Court concluded that the mere display of one's buttocks in public did not constitute indecent exposure but that the display of more private anatomical parts did constitute indecent exposure. In the finest traditions of the English common law, we now have a definition of what constitutes a nonpunishable "mooning."

## Miscellaneous

Courts are required to conduct their business in accordance with certain uniform rules in order to protect the constitutional rights of the parties to the cases. These rules are called *The Rules of Criminal Procedure* and *The Rules of Civil Procedure*. Civil procedure rules are used in civil cases and criminal procedure rules are followed in criminal cases. These rules not only protect the rights of the parties; they also assure a uniform way of doing business in the courts, thereby helping the system to run smoothly. Courts themselves will also establish what are called "local rules," which are created to further the efficient administration of justice in a particular court in a particular locality. The smart lawyer always investigates these local rules before proceeding too far in an unfamiliar court. These rules are complex and can be difficult to manage for someone not familiar with them, especially a layperson seeking to represent himself or herself.

Finally, there is the distinction between civil and criminal law. Criminal law is the name for a body of law in which crimes and their punishments are described and one party to the case is either the state of North Carolina (designated *State*) or the United States of America (designated *U.S.*). Civil law is the body of law describing rules which are related to noncriminal activity and which are intended to enable citizens to work and live together. Anyone can be a party to a civil case. Instead of fines or imprisonment as a penalty for losing (as in a criminal case), the loser in a civil case may be required to pay compensation to the winning parties, to do something, or to cease doing something.

# Chapter 3

# Creation and Organization of Fire Departments

## Introduction

North Carolina recognizes two basic types of fire departments: volunteer and paid (or career). For many reasons, especially during times of budgetary stress, these two types are combined within a single department. Typically, the personnel of municipal (paid/career) departments are augmented by volunteer firefighters, and the personnel of volunteer fire departments are augmented by paid personnel during specified periods of the work week. Departments utilizing combinations of paid/career and volunteer personnel are sometimes identified as "Combination" departments. A term frequently used in describing fire departments and defining their operational authority is "Authority-Having Jurisdiction" (AHJ). AHJ identifies the governmental authority which has jurisdiction over a particular geographical area or subject matter. For example, the Carteret County government is the AHJ for all unincorporated areas of Carteret County, N.C.; the Town of Newport government is the AHJ for all areas within the boundaries of the town and its extraterritorial jurisdiction (ETJ); and the Nuclear Regulatory Commission is the AHJ for matters relating to nuclear energy.

## Paid or Career Fire Departments

This text hereafter will use the term "paid" to designate firefighters and/or departments who are paid regular wages and salaries and departments whose staff are paid regular wages and salaries. NCGS 160A-291 permits, *but does not require*, a municipality to establish a fire department, appoint a fire chief, and employ additional firefighters. It also authorizes the municipality to prescribe the duties of the department. This statute gives a municipality wide discretion in the organizational structure of the department and assignment of duties. These are ordinarily accomplished by the enactment of a municipal ordinance which officially establishes the department, describes some of its duties, and sometimes indicates the organizational structure of the department. This type of enactment appears to be declining, however, because of the constantly evolving and increasing complexity of fire department operations. Enacting and/or amending ordinances can become administratively troublesome for a municipality, so the current trend appears to be to enact ordinances with general descriptions of structure and duties, leaving flexibility available to the chief officers and the municipal government to adapt the personnel and equipment of the department to evolving needs. In at least one instance, North Carolina's courts have ruled that a municipality can require its fire department to perform duties not directly associated with fire prevention or suppression, basing their decision on the language in the last sentence of NCGS 160A-291.

The duties of a municipal fire chief are found in NCGS 160A-291. In addition to any duties imposed by the municipality, the chief is responsible for the operations of the department on and off the fire ground, submission of annual reports to the governing body of the municipality, and, upon occasion, some building-code enforcement activities.

Personnel matters in municipal departments are governed by the North Carolina State Personnel Act, together with any personnel ordinances enacted by the AHJ. Matters of compensation are subject to the Fair Labor Standards Act (FLSA) and any municipal ordinances which are not in conflict with FLSA.

An emerging fire-protection entity in North Carolina is the county fire department. Firefighters in these departments can be paid or volunteer, although at present most of them are paid. Typically, such departments are appearing in counties where most of the geographical area is included within municipalities. New Hanover County, for example, is small and consists of the city of Wilmington, a few small municipalities, and a comparatively small, unincorporated area. It has organized a county fire department, consisting of paid personnel, augmented by volunteers. NCGS 153A-233 authorizes counties to organize and maintain fire departments, enter contracts with volunteer fire departments, and appropriate funds to engage in these activities. Otherwise, the description of authority is not unlike that set out in NCGS 160A-291. Once organized, a county fire department operates in much the same fashion as a municipal department.

Another aspect of county-related fire protection which ordinarily involves paid personnel is the office of the county fire marshal. NCGS 153A-234 authorizes a county to appoint a fire marshal and employ assistants for the fire marshal. The statute allows the county to assign whatever duties to the fire marshal the county chooses, including advising the county board on fire service-related matters, coordinating fire training and suppression activities which come under the county's control, coordinating fire-prevention activities which are under county control, providing assistance to volunteer fire departments within the county, and conducting fire inspections as authorized by the applicable building codes. Duties involving responses to natural and man-made disasters are also being handed directly to fire marshals, or sometimes indirectly to them by incorporating the fire marshal's office into the county's emergency management office.

## Volunteer Fire Departments

The term "volunteer fire department" can be somewhat confusing because there exist in North Carolina municipal fire departments staffed by volunteers. In these instances, the departments are in fact bona fide departments of municipal governments which happen to be staffed by unpaid personnel. For purposes of this chapter, a "volunteer fire department" is one which meets certain basic criteria specified in the North Carolina General Statutes.

The first of these criteria is that the department must be incorporated in North Carolina as a nonprofit corporation in accordance with Chapter 55A of the North Carolina General Statutes. Once incorporated, the department becomes a private corporation, although it provides a service to the public on behalf of a political subdivision of the state. The department must maintain fire apparatus of a value of at least $5,000 and keep records of its activities. Before it can deliver any services, it must enter a contractual relationship with some governmental unit (a political subdivision of the state) to deliver services within a geographical area under the jurisdiction of the political subdivision. Typically, this contractual relationship is represented by a written agreement signed by the department and the political subdivision, known as a "county contract" or a "municipal contract."

The document which creates the corporation is called its Articles of Incorporation. Sometimes it is referred to as the corporate charter. It is issued by the office of North Carolina's secretary of state after proper documentation has been submitted and requisite filing fees paid. It declares the corporation to be in existence and establishes its name. A properly incorporated volunteer fire department is, in reality, two entities operating under the authority created by the Articles of Incorporation. One entity is the larger organization, the corporation itself, which owns the assets of the corporation (land, buildings, equipment, etc.) and the other entity is the service-delivery agency of the corporation—the fire department.

The corporation consists of its board of directors, corporate officers, and members. The department consists of the persons who deliver the service—the firefighters. Each entity has separate responsibilities. The corporation, through the board of directors and corporate officers, establishes and implements policy in much the same manner as a municipal or county board. The department, through its command structure and members, delivers the services. In larger volunteer departments, it is easy to create two governing bodies, one for the parent corporation and a second one for service delivery in the form of the command structure and members of the fire department. Smaller departments, however, generally intermingle the two entities, with little distinction between corporate and fire department officers. This close intermingling and centralizing of authority can create an atmosphere vulnerable to abuse by the command structure as well as corporate misconduct. Given the increasing size and scrutiny of fire department budgets, it is advantageous to any department to utilize this supervisory separation whenever possible. As of the date of this manuscript, most scandals involving volunteer departments have involved money, with some perpetrators receiving serious felony convictions ! Under ideal circumstances, the department should have a board of directors, a set of corporate officers (for example, a president, a vice president, a secretary, a treasurer, and so forth) and a command structure consisting of a chief and such subordinate leadership as may be provided in the bylaws of the corporation and/or the operating guidelines of the department, with each layer in the structure staffed by a separate group of persons.

As a free-standing, independent, nonprofit corporation, the volunteer fire department (VFD) is permitted to establish its own bylaws, rules, regulations, or operating guidelines as long as these do not violate the law. Basic corporate activities are governed by the provisions of Chapter 55A and the federal and North Carolina revenue codes, while service delivery falls under a set of rules which will be discussed in more detail in following chapters of this text.

## Combination Departments

There is no statute which recognizes combination departments. They have resulted from necessity on an as-needed basis and in two configurations. The basic legal character of each organization is determined by the nature of its respective parent organization. Thus, a volunteer department which hires some paid staff remains a volunteer department; and a paid department which augments its staff with volunteers remains a paid department.

A VFD which has augmented its staff with paid personnel remains a free-standing, autonomous entity able to write its own rules unless specific laws exist which govern the activities of paid personnel (for example, OSHA), in which case the particular law or regulation must control the activities of the paid staff. This is an important issue which must be watched closely by the leadership of any VFD which utilizes paid staff.

Paid departments which augment their staffs with volunteers likewise have issues which they must address. Typically, the most difficult one is the method by which the volunteers are integrated into the department. The AHJ can integrate the volunteers into its personnel policy, which may not be the most appropriate method because issues relating to the volunteers now become the direct responsibility of the governing body of the AHJ. Another approach, requiring less involvement by the AHJ, is to allow the volunteers to form a separate nonprofit corporation or association and then contract with the AHJ to provide the services. A third alternative is for the AHJ to authorize the fire chief to solicit volunteers and integrate them into the department, utilizing the operating guidelines (SOGs), bylaws, or some other group of internal rules of the department. This alternative would allow the volunteers to retain their identity separate from the paid personnel and keep intact the cost-saving benefit which led to the involvement of the volunteers in the first place.

## The Role of the Volunteer

Given the heightened interest in emergency response and consequent increased demands for funding to enhance existing response capabilities, volunteers are becoming subject to both heavier demands for service and tighter scrutiny from the AHJs which provide their funding. The days of being able to provide adequate emergency response solely with donated funds are over. Volunteers must be prepared to deal with the requirements which accompany funding originating from AHJs. The AHJs, in turn, should understand the operations of paid and volunteer departments and thereby appreciate the extraordinary savings given to taxpayers by the presence of volunteers.

Deep intrusion into the VFD operations by AHJs creates all sorts of problems. It can establish an unintended employer/employee relationship between the VFD and the AHJ, with consequent regulatory-compliance expense to the taxpayers. The intrusion can be perceived by the VFD to be a direct assault upon its independence. The greatest strength, as well as the "Achilles' Heel," of the volunteer fire service is its sense of independence. This is well documented through the history of America's fire service, which began as an all-volunteer organization. Diminishing this sense of independence discourages volunteers and causes reductions in morale and staffing. The resulting loss of personnel inevitably leads to the hiring of paid personnel and significant cost to the taxpayers. There are VFDs in North Carolina today which provide adequate fire protection to their response districts for less money annually than it costs to hire four paid firefighters. Obviously, the savings to North Carolina taxpayers is enormous—in the hundreds of millions of dollars. Today, between 70 percent and 75 percent of North Carolina's fire protection is provided by volunteers.

## Chapter 4

# The Authority to Act

This chapter will discuss the authority of fire departments to deliver their services. The rules relating to issues concerning this authority are essentially the same for both paid departments and VFDs. North Carolina's courts long have recognized, at least indirectly, the authority of an emergency responder to perform its duties upon arrival at the scene to which it has been summoned. As will be seen later in the materials relating to civil liability, departments, in at least one instance, have been held obligated to provide assistance once they announce their intent to respond.

North Carolina is one of a few states whose fire departments have been given a very broad statutory authority to engage in fire suppression and fire-protection activities. NCGS 58-82-1 empowers fire departments to take all reasonable steps to extinguish fires and protect life and property from fire. It also provides a mechanism for dealing with anyone who interferes with a department engaged in fire suppression or other protective, fire-related activities.

However, other activities, such as rescue services and emergency medical services (EMS), are not so clearly defined by statute. For instance, while EMS may be a part of the duties of a particular department, the authority to deliver those services does not exist in NCGS 58-82-1 because the statute deals only with fire-related activities. The authority to deliver EMS services is granted to the AHJ and its fire department and arises as a result of the decision of the AHJ and not as the result of a specific statute. However, once an EMS activity is initiated, the responding department becomes responsible for the consequences of its actions, as will be explained later in this chapter. EMS is an acquired responsibility for a department, although with the increasing emphasis upon unified service-delivery systems and first responders in many jurisdictions, AHJs are adding EMS as a standard component of their fire departments' tables of organization. The authority to act specified in NCGS 58-

82-1 has not yet been used to challenge the EMS activities of a department, but, given the apparent fondness of the American public for litigation, the fire service should be prepared to deal with litigation in the future unless the General Assembly amends existing statutes or enacts others to make the law more consistent with the duties imposed upon fire departments.

Hazardous Materials (HAZMAT) is another response activity where the rules are imprecise. When North Carolina and federal authorities originally recognized the need for a comprehensive system of response to incidents involving HAZMAT, the North Carolina State Highway Patrol was handed primary responsibility for response management because the primary route for transporting HAZMAT in North Carolina has always been its system of highways. However, it became apparent almost immediately that other agencies would need to participate in the response process because the highway patrol was neither trained nor equipped to handle such responses. The focus of its duties was, and remains, law enforcement, as it should be. This left North Carolina with a need for a statewide HAZMAT response system but without an agency to manage the responses. The only readily available agency was the fire service. Federal law mandated a response system pursuant to 20 CFR 1910, and thus the role of the fire service in HAZMAT was born.

In 1977, the general assembly enacted Chapter 166A of the General Statutes, the North Carolina Emergency Management Act. It consists of four articles and covers topics related to emergency management, hazardous materials response, volunteer leave, and an interstate compact to provide assistance to other states who are members. The act establishes a statewide system for emergency response to disasters or major emergencies. It places the governor in overall command, with the Department of Public Safety as the lead agency in implementing the response. The act further provides for local government agencies to implement responses at the local level (see NCGS 166A-5). At the local level, in accordance with the protocols developed there, the fire service becomes involved in the multifaceted aspects of emergency management and acquires its authority to act.

Also found in the act is NCGS 166A-20 (and the statutes following) which creates the HAZMAT regional response program. The act instructs the Department of Public safety to establish no fewer than six regional teams to respond to HAZMAT emergencies. The secretary of the Department of Public Safety is empowered to establish rules for the HAZMAT teams, and through these rules fire departments become the lead participants in the emergency response teams and acquire their authority to act.

In Chapter 130A of the General Statutes, "Public Health," the secretary of Health and Human Services is authorized to take steps to protect the public from hazards to its well-being. In NCGS 130A-6, the secretary is authorized to delegate his or her authority to the local level in order to perform these functions. Once again, fire departments become parties to local protocols relating to the abatement of hazards to the public's health and well-being and acquire their authority to act. NCGS 130A-17 establishes a right of entry in order to enforce the necessary rules. However, the entry must still be undertaken in accordance with existing constitutional rules relating to search and seizure. NCGS 130A-2 defines an "imminent hazard" as a situation likely to cause a threat to human life, a threat of serious physical injury, an immediate threat of adverse health effects, or a threat of irreparable environmental damage if no immediate action is taken. Utilizing available resources and after notifying or attempting to notify the owner of the instrumentality creating the imminent hazard, a local health department possessing the authority of this statute may enter any property and undertake corrective action. In this circumstance, when it responds, the fire department does not act on its own but works under the authority of the local health department and any existing protocols.

Environmental considerations are now important concerns for fire departments when planning live-burn exercises. The rules dealing with these considerations have their origins in Chapter 130A. NCGS 130A-444 deals with asbestos inspections and disposal, and NCGS 130A-453.01 begins the section of the statutes which discusses lead-based paint. We in the fire service would do well to follow these rules in order to avoid possible penalties for violating them, especially when dealing with acquired structures in live burn exercises.

Occasionally, in the course of responding to an incident, a department may wish to detain one or more persons in order to discuss some aspect of the incident. Not even NCGS 58-82-1, with its broad grant of authority, allows a fire department to arrest or detain someone. However, in Chapter 15A of the General Statutes we find two statutes which provide some assistance. NCGS 15A-404 prohibits a private person (non-law enforcement officer) from arresting someone. In other words, North Carolina does not allow "citizen's arrests." However, NCGS 15A-404 does allow a private person—for example, a firefighter—to *detain* someone if the firefighter has probable cause to believe the detainee committed, in the presence of the firefighter, a felony, a breach of the peace, or a crime involving physical injury to another person or theft of or damage to property. Once detained, the detainee must be turned over to the custody of law enforcement as soon as possible. NCGS 15A-405 authorizes law enforcement officers to enlist the assistance of non-law enforcement personnel to make arrests or prevent the escape of persons who have been arrested. The law enforcement officer must make the request in order to establish the lawfulness of the activity. Once so involved, the non-law enforcement personnel are entitled to the same benefits and protections afforded law enforcement officers should they be injured in the performance of these duties or be accused of misconduct.

# Chapter 5

# Investigations and Inspections

## Basics

This chapter will discuss two important duties assigned to fire departments which, until recently, were often regarded as secondary to the department's primary duties of fire suppression and fire prevention: determining the cause and origin of fires, and fire-code enforcement. The duty of a department to investigate the cause and origin (C&O) of a fire has existed since 1899, when the first version of NCGS 58-79-1 was enacted by the legislature as part of Chapter 69 of the General Statutes. When North Carolina incorporated a fire prevention safety code into its building code in April 1992, fire departments which in the past had performed fire safety inspections utilizing nonuniform standards suited to the needs and abilities of individual departments and AHJs discovered that the rules had changed radically and that now the mission was to be accomplished by personnel utilizing a uniform set of rules and standards as specified by the state's building code. The organization of the code is such that fire inspectors need not be members of the fire service. They need only complete the qualification requirements specified in Title 11, Chapter 8 of the NCAC. Recognizing the advantage to be afforded the taxpayers of North Carolina by cross-training existing personnel, many AHJs have moved forward with aggressive training programs to train firefighters as code enforcement officials (CEOs) to enforce the fire safety portions of the Building Code.

C&O investigations and code enforcement activities require an action subject to close scrutiny under the law: entry into the property of another person without the consent of that person. This action, under ordinary circumstances, is regarded, at the very least, as an invasion of privacy. The Fourth Amendment of the Constitution of the United States and Article I, Section 20 of North Carolina's constitution are intended to establish and protect this privacy. Both constitutions also provide for the lawful invasion of the privacy of a person if the entry is conducted pursuant to certain standards or the entry is conducted with the consent of the owner or occupant of the premises in question.

Both C&O investigations and code enforcement activities are dependent upon the acquisition of information for their effectiveness. This information is acquired through a search conducted by either the fire-cause-and-origin investigator (FI) or the CEO and is called an *administrative search* because it is undertaken in pursuit of information that does not ordinarily lead to criminal prosecution. The law specifically distinguishes it from a search undertaken in pursuit of evidence of criminal activity in NCGS 15-27.2(g).

Many of the basic rules relating to administrative searches were established by two cases from the United States Supreme Court: *Michigan v. Tyler* and *Michigan v. Clifford*. Both cases involved searches which were incident to suspicious fires and which were undertaken by fire departments with the assistance of law enforcement agencies.

## *Michigan v. Tyler* 436 U.S. 499 (1978)

A fire was discovered in a commercial building sometime prior to 2:00 a.m. on 21 January 1970. By 2:00 a.m. the fire department was on the scene and the chief had declared the fire to be of a suspicious nature. Fuel containers were discovered and the interior of the building was searched. At 3:30 a.m., a police detective arrived and joined the investigative team. At 4:00 a.m., the weather and poor visibility caused the fire department to take up and depart, followed almost immediately by the detective. At 8:00 a.m. the fire department returned and the fire investigation began in earnest. At 9:00 a.m., the detective returned and rejoined the investigation. At no time during 21 January 1970 was any type of search warrant or consent from the owner of the property acquired by the investigators. On 16 February 1970, a Michigan State Police investigator visited the abandoned scene without a search warrant and retrieved additional evidence of arson. Criminal charges were brought against the owner of the business and pursued through the Michigan courts. The highest appellate court in Michigan then ruled that all the evidence obtained subsequent to 4:00 a.m. on 21 January 1970 was inadmissible in court because the searches by which it was obtained violated the Fourth Amendment. The case was then appealed to the United States Supreme Court.

The Court first ruled that the Fourth Amendment applies to firefighters in the performance of their duties and includes both administrative and criminal searches. The Court then acknowledged that an emergency is a circumstance in which authorities can enter property without a warrant or the consent of the owner/occupant of the property and that a fire is clearly an emergency. The Court pointed out that once emergency responders enter the scene of an emergency and initiate activities to cope with the emergency, the responders, in effect, stand in the shoes of the owner/occupant and can authorize entry to the scene to such persons or agencies as the responders perceive to be necessary to assist them with their response. The Court held that a firefighter, when engaged in the performance of his duties in response to an emergency, can lawfully seize evidence of criminal activity—for example, arson—if the evidence is in plain view of the firefighter. The Court specifically rejected the idea that final suppression of a fire terminates the authority of a fire department to remain on the scene. The Court also stated in its ruling that a C&O investigation clearly serves a compelling public interest and is part of the response to the fire-created emergency. The Court further held that a warrant is not necessary if a search is conducted within a reasonable time after the fire has been suppressed and that the search is, in fact, a continuation of the response to the existing emergency. Specifically in *Tyler*, the Court found that the departure and return of the department (and other investigators) was reasonable, especially considering the weather conditions, and that all evidence gathered on 21 January 1970 was admissible in court. The Court held, however, that the warrantless entry by the Michigan State Police investigator required a criminal search warrant and that all evidence gathered on 16 February 1970 was not admissible in court.

## *Michigan v. Clifford* 464 U.S. 287 (1984)

At 5:40 a.m. on 18 October 1980, the fire department arrived at a fire in a residential dwelling. By 7:04 a.m., the fire was extinguished, and the fire department took up and, along with the police department, departed. During the initial attack on the fire, a firefighter discovered a container later determined to contain an accelerant within the structure and removed it to the paved driveway that served the structure, where it remained after the departures of the fire and police departments. At 1:00 p.m. on the same day, the fire and police department investigators returned to the scene, *without search warrants or con-*

*sent of the owner*, and entered the premises, initiating a search for evidence of the cause and origin of the fire and of possible arson. Upon their arrival, investigators discovered workmen, employed by the owner, pumping water out of the basement and attaching plywood to cover the windows broken in the course of the earlier suppression activities. By 1:30 p.m., the search was underway, and the container left by the firefighters in the driveway was collected as evidence of possible arson. Upon entry of the premises, the investigators discovered a point of origin in a stairway, accompanied by a distinct odor of accelerant. In the kitchen, appliances were discovered to have been modified, and the team proceeded to go throughout the house, searching for additional evidence of arson. Subsequently, the case worked its way through the Michigan courts and arrived at the United States Supreme Court.

The primary question presented to the Court was whether or not a firefighter, in the absence of exigent circumstances (an emergency), could enter a dwelling without a search warrant or consent of the owner/occupant thereof to investigate a recent fire. Unfortunately, the Court did not enter a majority opinion. It entered a plurality opinion, with four justices affirming, one justice concurring (Justice Stevens), and four justices dissenting, leaving us with guidance but no final ruling. The plurality of justices held that administrative searches of the type undertaken in this instance must be conducted in accordance with the Fourth Amendment and listed three considerations for warrantless, nonconsensual searches. First, are there legitimate privacy interests in a fire-damaged property which must be protected by the Fourth Amendment? Second, do exigent circumstances justify entry without a warrant or consent? Third, is the object of the search an administrative or criminal investigation?

The plurality stated that a fire is an exigent circumstance and that once the response is initiated the firefighters may remain for a reasonable time to conduct an investigation. A C&O investigation serves a compelling public interest and no warrant should be required at this point in time. However, the plurality stated that once the emergency responders leave the scene, they should not reenter the premises without a warrant or the consent of the owner/occupant. If the search is C&O, an administrative search warrant is appropriate. However, such a warrant does not entitle the searcher to wander throughout the premises at random. If the search is conducted for evidence of criminal activity, a criminal search warrant is necessary, absent the consent of the owner/occupant of the premises.

Justice Stevens, in his concurring opinion, stated that firefighters have an absolute right of entry in response to an emergency and the right to remain as long as necessary thereafter to conduct an investigation. He opined also that any search outside the area of origin of the fire should be conducted with either a warrant or consent of the owner/occupant.

Several factors illustrate the difference between *Clifford* and *Tyler*. First, the makeup of the Supreme Court had changed in the six years between the two cases (different justices served on the Court when *Clifford* was written and the search was not the continuation of an earlier search as it had been in *Tyler*). Second, the investigators clearly had interrupted the initial investigation and begun a fresh one. Third, it was obvious that by the time the investigators in *Clifford* had returned, the owner of the property had reasserted his control over the premises and was taking steps to secure his property by pumping out the water and covering the windows. Finally, the property in *Clifford* was a residential dwelling rather than a business establishment. Courts have long recognized that a merchant has a lower expectation of privacy than a private homeowner because the merchant generally encourages unrestricted entry into his premises in order to generate business, while the homeowner lives in a fashion which is intended specifically to protect his privacy.

So where does all this leave fire officials? It is clear that they can conduct any type of search with the consent of the owner/occupant. The response to an emergency will get the fire department on the premises to deal with the emergency and to conduct a C&O investigation without a warrant or consent. However, once the emergency has been dealt with and control of the scene or premises has been returned to or reclaimed by the owner/occupant, any subsequent return to the scene/premises must be accomplished with the consent of the owner/occupant or a search warrant. When evidence of criminal activity is discovered in the course of a C&O investigation, it is advisable to secure the premises, notify appropriate law enforcement agencies, and allow them to proceed as they deem appropriate. In effect what has occurred is that an administrative investigation has transitioned into a criminal investigation. This leaves the C&O investigator and the responsible fire department in a bit of a quandary because they still have an obligation to complete a C&O investigation, but law enforcement authorities have an equally important obligation to preserve and investigate the crime scene. At this point, cooperation between the agencies involved is a necessity. Some sort of working relationship whereby the fire investigators can complete their C&O investigation while law enforcement authorities pursue their criminal inquiry and neither group interferes with the other's activities. Some preplanning involving the fire department and the AHJ's law enforcement investigators might facilitate this process. Because each incident scene is unique, a general framework for tackling the problem likely is best one can accomplish, but it would be an important step forward.

The Court in *Clifford* did leave fire officials with some additional guidance regarding the information required to identify the need for an administrative search warrant in the course of a C&O investigation. The plurality identified information which should be presented in the application for the warrant (the affidavit). The affidavit should state that a fire of unknown cause and origin has occurred on the property in question. The scope of the search (the area and items to be examined) should be identified and should include things being sought by the investigators. Also, the search should be conducted in a fashion to avoid, as much as possible, undue intrusion upon the privacy of the owner/occupant. In addition, the search must be conducted at a reasonable and convenient time. The Court also offered some guidance regarding the search itself. Evidence of arson in plain view may be seized by the investigator. However, an administrative search is not a license for the fire officials conducting the investigation to roam freely through a private residence. If evidence and adequate probable cause of criminal activity are discovered, the C&O activities should be stopped and a criminal search warrant should be acquired by appropriate criminal investigative authorities. In its opinion, the Court identified the point of transition from a C&O investigation to a criminal investigation and offered guidance in the management of the situation.

In summary, the plurality opinion left some guidance in the management of C&O searches when it becomes necessary to abandon the scene to the owner/occupant prior to the completion of the investigation. Fire officials can return and investigate without a warrant with the consent of the owner/occupant or if another emergency (typically a rekindle) occurs at the scene. Otherwise, they should obtain the warrant.

## North Carolina

The North Carolina rules regarding administrative searches have built upon the principles set out in *Clifford* and *Tyler*, and a statute, NCGS 15-27.2. In addition, North Carolina's appellate courts have issued some opinions regarding administrative searches which augment the United States Supreme Court cases and the statute. Most of these cases have arisen as a result of actions undertaken by North Carolina's Occupational Safety and Health Administration (OSHA). Ordinarily, the consequence of obtaining evidence as the result of an entry without the consent of the owner/occupant or a properly prepared and obtained search warrant is that the evidence so obtained will be inadmissible in most courts or administrative hearings as "fruit of the poisoned tree" because it was not obtained utilizing proper constitutional procedures. However, if the conduct surrounding the search is sufficiently outrageous, the searchers themselves, together with their employers, may find themselves subject to criminal or civil liability.

# NCGS 15-27.2 Administrative Search and Inspection Warrants

The statute authorizes officials of state or local government to conduct searches or inspections with warrants pursuant to the statute with or without the consent of the party whose property is being searched, as long as the particular activity is one otherwise authorized by North Carolina law. The statute supplies the guidelines for such a warrant when one is necessary.

The warrant may be issued by the following officials of the general court of justice: a judge, a magistrate, a clerk of superior court, an assistant clerk of superior court, or a deputy clerk of superior court whose respective territorial jurisdiction includes the area to be inspected or investigated. This means that the application for the warrant should be submitted and the warrant obtained in the courthouse with jurisdiction over the property in question. For example, a fire investigator working a fire in Carteret County could not obtain a warrant in Onslow County to search that property. He or she would have to obtain the warrant in Carteret County.

Before the issuing officer can issue the warrant, the issuing officer must be satisfied that certain conditions are met. This matter of "satisfaction" can, upon occasion, be a subjective one—that is, it can occur at the discretion of the issuing official. This is an important consideration for the inspector/investigator seeking the warrant. What may be satisfactory to one judicial official may not be satisfactory to another. The law enforcement community has been aware of this discretionary practice for many years and always attempts to find officials issuing its warrants who appear to be sympathetic to the needs of the moment. The subjectivity of an issuing official should be a consideration for inspectors and investigators as well. Administrative Search Warrants are seldom seen in some jurisdictions, so it may be necessary for the inspector or investigator to educate an issuing official unfamiliar with the procedure. This can be accomplished by the investigator or inspector himself or herself or by the attorney for the AHJ who accompanies the person applying for the warrant. The secret for success is to be well prepared.

The first requirement to be satisfied is to demonstrate to the issuing official that the place to be inspected or investigated (searched) is within the jurisdiction of the issuing official and that the intended search is either part of a legally authorized program of inspection or investigation which naturally includes the property in question or that there is probable cause to believe that a particular condition, object, activity, or circumstance exists which makes the search necessary. This information must be disclosed in an affidavit submitted by the applicant for the warrant, executed under oath or affirmation. The second requirement is that the issuing official must examine the applicant under oath or affirmation regarding the circumstances which led to the need for the warrant.

Next, assuming that the issuing official is satisfied with the affidavit and its supporting information, the warrant is issued to the applicant. The warrant is subject to certain time limitations, however. A search warrant is not a "license to run wild." It is signed by the issuing official and marked with the date and time of issuance. A warrant for basic code enforcement activity (such as a fire inspection) is good for twenty-four hours from the date and time it is issued, can only be served (executed) between 8:00 a.m. and 8:00 p.m., and must be returned for final signature by an official with authority to issue it within forty-eight hours of date and time of issuance. A C&O investigation warrant is good for forty-eight hours from date and time of issuance, may be served at any time, and must be returned for a final signature within a reasonable period of time after its execution or at the expiration of the forty-eight-hour period if it is not used. Adherence to the time frames in the statute is *vital.* Some lawyers make very good livings defeating search warrants of all types by discovering and taking advantage of technical errors involving the issuance or execution of a warrant, or technical errors on the paperwork itself.

The statute requires that the warrant be served upon the owner or possessor of the property in question. Confusion often arises when tenants occupy portions of a property which must be searched pursuant to a warrant. The lessor (landlord or landlady) *cannot* consent to the search of his or her tenant's property. The courts have long held that the execution of a valid lease, written or oral, delivers the custody of the property in question to the tenant, to the exclusion of the lessor, for the duration of the lease. This means that reasonable efforts must be made by the searcher to notify the tenant whose property is the subject of the impending search.

The exact procedure for serving the warrant does not appear to have been explained precisely by the courts. The traditional method, however, is to approach the person upon whom the warrant is to be served, read both the affidavit and the warrant itself to that person, and then proceed with the search. If the property is vacant and can be entered without the use of force and if the owner or possessor cannot be located after a duly diligent attempt to locate the owner or possessor, the searcher may serve the property in question and conduct the search. Serving the property consists of reading the appropriate documents *out loud* to the property to be searched and then attaching a copy of the warrant to the property. The law does not state how quickly or how loudly the documents must be read—only that such activities take place prior to entry by the searchers. The prudent investigator would do well to document all attempts to locate owners or possessors prior to the search because if the procedure is attacked at a later date, one of the issues to be raised will concern efforts to contact these parties.

The statute also places limitations on the types of evidence which can be collected and utilized in the course of an administrative search. Only "a condition, object, activity, or circumstance which it was the legal purpose of the search or inspection to discover" is admissible in any sort of legal proceeding or is useable for the purpose of generating probable cause for further warrants, unless the warrant was not constitutionally required in the first place. Arguably, an entry made with the consent of the owner or possessor of the property can allow developed information which falls outside the boundaries of the authority of the searcher to be admitted in certain future proceedings as long as the court can be satisfied that the searchers did not abuse their authority. This argument sits on shaky legal ground, to say the least, especially when examined in light of *Clifford* and *Tyler.*

Who should serve the warrant? A basic principle of North Carolina search warrant law requires that the person who serves the warrant conduct the search. This does not prohibit the server of the warrant from enlisting the help of others to assist in the search as long as the assistance provided is reasonable in the eyes of the court. It is not uncommon for a searcher to be accompanied by one or more law enforcement officers, either to assist in the search or to provide security. If law enforcement officers are present, it is vital to remind them that the search is an administrative search and very different from a search for evidence of criminal activity. *Clifford* and *Tyler* suggest rather strongly that if the administrative search implies the presence of criminal activity, a criminal search warrant should be obtained to investigate the circumstances in question.

Another basic rule of searches is that property may be searched at any time with the consent of the owner or possessor. Consent carries two limitations, however. First, consent must be given by the proper party—that is, someone with the authority to do so. The classic example is the lessor (landlord or landlady). As mentioned earlier, a lessor cannot consent to the search of the premises of a tenant. The lessor can consent to the search only of premises over which he or she has dominion and control. In the case of an apartment building, for instance, the lessor can consent to searches of common areas (stairwells, lobbies, recreation areas, passageways) or any apartments not leased to tenants. This limited form of consent requires that the searcher identify as thoroughly as possible the individual units to be searched and, if possible, the identities of the tenants who signed the leases in order to give them notice of the impending search. The second problem with consent is that it is revocable at any time. The protections of our rights of privacy are powerful and are imposed by statute, constitutional law, and the courts. It is a well-established rule that consent may be revoked at any time, whether the consent is written or oral. The searcher confronted with a revocation must depart immediately or face the possibility of accusations of trespass or similar crimes and any evidence gathered after the revocation is not admissible in a court or other proceeding. In addition, the courts do not allow a searcher to use intimidation or subterfuge to get around consent. In a case from Illinois, the United States Supreme Court made it abundantly clear that such behavior is unacceptable. In the world of consent, "No!" means "No!"

NCGS 15-27.2 also requires that the property to be searched be described adequately. Typically, this occurs first in the affidavit and then in the warrant. If the property is not described adequately, the warrant could be invalidated and all evidence gathered as a result declared inadmissible. There are many ways to describe property, and the most commonly accepted method is to use a combination of the address, the exterior description—color of roof, shutters, siding, and so forth—and the basic shape and materials of the structure. A totally unacceptable method to describe a property is to identify it in the affidavit and warrant as something to be described at a later date. If there are apartments or other units inside the primary structure, these must be described as well if they are to be searched. The amount of description necessary for properly examining individual units is not clearly defined, so its sufficiency in any given instance will probably be left to the discretion of the judge before whom the search warrant is presented.

The affidavit must be attached to the search warrant and both instruments served upon the persons whose property is to be examined. The Administrative Office of the Courts (AOC) has simplified the process somewhat by developing two administrative search warrant forms which incorporate all necessary information required by statute. One side of each form contains the affidavit and the other contains the warrant itself. If one plays by the rules, is careful in completing the form, and follows the required time frames, the search warrant process can be made both efficient and effective. All AOC forms are identified by numbers, and using these numbers makes the process of locating the forms much easier for both the applicants for the warrants and the courthouse personnel from whom the applicant seeks the forms. These forms now are available online. Enter "NCAOC" in the search engine and locate the section dealing with forms and follow the instructions. Enter the form number, click on search, and a form should appear in either PDF or Word format. Print the blank form. The current numbers are AOC-CR-913M and AOC-CR-914M for each form, respectively. One form is used when the search is pursuant to a system of periodic inspection/investigation, and the other form is used when a particular circumstance, including a fire, is to be inspected or investigated.

## Fire Investigations

Fire investigations are ordinarily undertaken in two contexts—either to determine the cause and origin of a fire or to develop information to identify and convict a fire-setter. The C&O investigation is conducted in a noncriminal context, while the investigation of a fire determined to be incendiary is conducted under the same basic rules as any other criminal investigation. Most arson investigations in North Carolina are undertaken by law enforcement officers with the assistance of fire service personnel, with law enforcement officers acting as lead investigators. This section will emphasize the rules applicable to C&O investigations, which are normally conducted by fire service personnel. It is not uncommon for a C&O investigation to develop evidence which indicates incendiary activity, resulting in the involvement of law enforcement investigators and the continuing involvement of some fire service personnel for some time after the incendiary determination occurs. The prudent C&O investigator should follow the recommendations of law enforcement personnel at this time in order to avoid possibly damaging the effectiveness of the criminal investigation. This transition from a C&O investigation to a criminal investigation was discussed by the Court in *Clifford* with a recommendation that criminal investigative rules should be applied to the inquiry once the transition begins.

## Cause and Origin (C&O) Investigations

The constitutional oversights for all C&O investigations are the Fourth Amendment and Article I, Section 20 of the North Carolina Constitution. NCGS 58-79-1 authorizes the Attorney General, through the SBI and, in the case of a municipality, through the fire chief or the police chief, and the case of a county, through the fire chief, the fire marshal, or the county sheriff, to conduct C&O investigations of all fires in which property is damaged or destroyed in order to determine whether the fire was caused by incendiary activity or carelessness. The investigation must be initiated within three days of the date of the fire (excluding Sundays) and may be supervised and directed by the SBI if the Attorney General decides such a procedure is appropriate under the existing circumstances. The investigator must, within one week of initiating the investigation, furnish the Attorney General, or his designee, a preliminary written report of facts relating to the cause and origin of the fire, together with such other information as may be requested. Currently, this information is collected on forms provided by the Office of the State Fire Marshal (OSFM) to all fire departments in North Carolina and the information and the information transmitted to OSFM through computer-based data processing. A few departments still furnish it in paper form. The reports must be maintained by the Attorney General, or his designee, and made open to public inspection. In many counties, the designee of the Attorney General for purposes of maintaining this information is the office of the county fire marshal. One problem yet to be addressed is the effect of recent privacy laws restricting access to personal information of United States residents on the apparently unrestricted release of information concerning C&O investigations. Frequently (and most certainly if the investigator is thorough), the information gathered in the course of a C&O investigation will contain personal information about the victim(s) of the fire which would be subject to the protections of the Privacy Act. The author suggests that the custodians of the investigative information, especially the fire department conducting the investigation, should regard such information as confidential and either redact the information or refer requests for its release to the Attorney General or an appropriate designee. As a practical matter, redacting the information and then delivering the report may be the least troublesome alternative.

If a C&O investigation is referred to the Attorney General under NCGS 58-79-1, the Attorney General, or his designee, may conduct further investigations pursuant to NCGS 58-79-5 to examine in greater depth whether the fire in question was the result of carelessness or incendiary activity. If a determination of incendiary activity is made, the Attorney General, or his designee, is directed to furnish such information to the appropriate authority for prosecution in criminal court.

NCGS 58-79-10 gives these investigators a broad grant of authority to conduct their inquiries. The designee of the Attorney General is authorized to issue summonses to witnesses to appear and testify under oath regarding the matters under investigation. The statute likewise authorizes the Attorney General's designees to enter the property under investigation at any time of the day or night to conduct their investigations. (This provision is subject to the holdings in *Clifford* and *Tyler and other constitutional restraints.*) The statute also imposes a strict duty of privacy and confidentiality upon the investigators regarding their findings and their witnesses.

Insurance companies will investigate a fire if it involves substantial loss or suspicious circumstances. This is especially true during difficult economic times, when insureds sometimes attempt to escape financial obligations by destroying their insured property and then hoping the insurance company will pay off the debt secured by a lien on the destroyed property. NCGS 58-79-40 obligates an insurance company to furnish certain information relevant to such investigations to C&O investigators, including the SBI. The materials to be furnished include insurance policies, policy premium payment records, claims history of the insured, and any other information or materials relevant to the investigation of the loss. NCGS 58-79-40(b) requires that the insurance company furnish the SBI with all information relevant to the investigation if the fire is determined by the company to be of suspicious origin. Subsection (c) of the statute provides protection from civil or criminal liability for investigators as long as the statements and other information furnished pursuant to the investigation were made and acquired without fraud or malice. This part of the statute appears to encourage the investigators to keep open minds and discuss theories of the case freely among themselves. The statute also requires the recipient of information from an insurance company to keep the information confidential until its release is required by legal proceedings.

Another aspect of fire investigation is the report generated by the department involved in the incident. Currently, there are two versions of these reports. The first is mandated by NCGS 58-79-45. The statute requires that whenever a department responds to a fire, the chief shall cause an incident report to be prepared and submitted on a form approved by the Department of Insurance. Currently, these reports are submitted using specified formats and procedures and may be submitted as hard copy or by electronic media. The original of the report is to be kept by the department, with an additional copy submitted to the local county fire marshal, or in the absence of the county fire marshal, to be county commissioners, for storage and future reference. The statute requires that the department and the county retain the reports for at least five years. The prudent department should retain its copy of the report for as long as possible. Frequently asked, but apparently unanswered, is the question: Who is entitled to a copy of the report held by the department? The statute does not offer any guidance. If an insurance claim is involved, the insurance company and the insured will usually seek copies from the department. Given the requirements of NCGS 58-79-40, it seems reasonable for the department for furnish a copy of the report to the insurance company and, possibly, the insured. However, beyond those considerations, a prudent department should refer any other requests to the fire marshal or county commissioners, pursuant to NCGS 58-79-45(c).

The second type of report is the department's internal incident report. Unlike the state-mandated form, there is no prescribed format for this one. It can be developed by the department for its internal use and ordinarily contains a narrative of the incident. With this form, the C&O investigator is able to document interviews and discuss theories regarding the incident. This report often contains sensitive information which should be considered confidential, its release permitted only when there is a need to know. The privacy of the victims of the fire and the confidentiality of the investigation are at stake. A good place to address these issues is the department's SOGs.

North Carolina law does not offer much specific guidance regarding C&O investigations, and, given the almost infinite variety of circumstances which can be encountered by the investigator, this seems appropriate. Other than Fourth Amendment and basic evidence preservation procedures, the investigator is left with his or her training and experience as guideposts. Because North Carolina has an evidence code which is extensive, complex, and beyond the scope of this text, specific questions regarding evidence should be referred to legal counsel. However, there are excellent courses in investigative techniques available through North Carolina's community college system, the Department of Insurance, the National Fire Academy, and other agencies. One readily available set of investigative standards are NFPA 921 and 1003, which are favored by North Carolina's courts and fire officials. They were created as nationwide guidance for investigators and are an integral part of many courses dealing with investigative techniques. This lack of uniform standards for investigations causes courts to turn to such sources as NFPA 921 and 1003 when they are attempting to determine whether or not an investigation has been conducted properly. A prudent investigator would do well to pay attention to such standards as those in NFPA 921and 1003.

## Investigation of Incendiary Events

The investigation of suspicious or incendiary fires or other suspicious events involves many of the same techniques as the investigation of nonincendiary events, but the rules regarding search warrants and interrogation of witnesses are radically different. In an arson investigation, the object of the investigation is not only the determination of C&O but also the identification and criminal conviction of a perpetrator. Because criminal penalties are at stake in these investigations, the courts have imposed strict standards of conduct upon the investigators, necessitating formal training and certification. The Department of Insurance established one level of certification—Certified Fire/Arson Investigator (CFI). Another is certification as a law enforcement officer by North Carolina. The CFI does not have arrest authority, as does a law enforcement officer, but may participate actively in these investigations, especially in circumstances where law enforcement authorities may lack some of the fire-investigative skills of the CFI. The CFI who participates in a criminal investigation must understand that the requirements of a *Miranda* warning may apply to questioning and that the custody of evidence and the acquisition and execution of search warrants will be subject to a set of rules very different from those which characterize a basic C&O investigation. It would be prudent for the CFI to follow law enforcement's lead in these matters.

The principles set forth in *Miranda v. Arizona*, which deals with verbal or written testimony only, are triggered when a suspect is in a "custodial" situation. In simplified terms, the suspect must be under the control of a law enforcement officer, or the suspect must reasonably believe that he or she is under the control of a law enforcement officer. The phrase "under the control of" can mean anything from being handcuffed in the back seat of a vehicle to being told to sit down in a room in a law enforcement facility and not to leave. Unfortunately, there are few hard and fast rules regarding what a custodial situation is and what it is not. Frequently, the courts must examine the facts of each situation to determine whether or not the *Miranda* warning was necessary. Experience has shown that subjects usually become more cautious regarding their statements once the warning has been given; thus it behooves the investigator to question a subject as much as possible in a noncustodial, nonthreatening context. Once the custodial situation has been established, the subject must be informed that he or she has the right to remain silent, that anything said subsequent to the administration of the warning can and will be used against the suspect in court, that the suspect has the right to legal counsel, and that if the suspect cannot afford counsel, one will be appointed to represent the suspect. Failure to administer the warning once the custodial situation has been established merely prevents any statements made by the suspect from being admitted into evidence against the suspect. Statements made prior to the creation of the custodial situation remain admissible.

Collection and preservation of physical evidence will be subject to closer scrutiny in an incendiary investigation than in a nonincendiary investigation. Documenting the chain of custody of the evidence and assuring that it is being stored in circumstances where the likelihood of tampering is minimal are vital considerations. Failing to follow certain basic rules of evidence management once the evidence has been acquired can make it inadmissible in court. One important rule which has emerged in the United States courts and which is being applied with increasing frequency in North Carolina's courts is the required use of the trial judge as a screener of evidence before it is presented in court. This role is described as that of a "gatekeeper," who prevents the admission of scientifically inaccurate evidence. The concept was introduced by the federal appellate courts in a series of appeals of criminal convictions for various types of unlawful burnings. The duty imposed upon the trial judge is to question a witness who intends to present the scientific evidence to determine whether or not the scientific theory behind the testimony is legitimate. This gatekeeping role becomes a vital consideration when an investigator presents an opinion regarding the C&O of the fire or other matters relating to fire behavior. Thus, it becomes important for the witness to be prepared to present physical evidence and also demonstrate to the trial judge that any theories regarding the behavior of the fire are based upon valid, researched scientific principles.

The basic rules relating to criminal search warrants are found in Article 11 of Chapter 15A of the General Statutes and are considerably more rigorous than those found in NCGS 15-27.2. The Supreme Court in *Tyler* and *Clifford* recognized the distinction and so have North Carolina courts. For instance, demonstrating adequate probable cause for the search is subject to a much more elaborate and demanding set of requirements in a criminal warrant than in an administrative warrant. Also, the areas to be searched and items to be seized must be described in the warrant very carefully. The investigator seeking to conduct a criminal search without the benefit of the subject's consent would be well served to consult with law enforcement authorities, including the staff of the district attorney, before seeking the search warrant. One cannot use trickery, subterfuge, or intimidation to induce consent, but North Carolina's appellate courts have approved the action of tricking a criminal into revealing sufficient information to generate probable cause for a search warrant. In one case, law enforcement officers tricked some drug dealers into opening a door, at which time the officers observed suspected drugs assembled on a table in front of the opened door. The court held that the opening of the door was not a violation of the criminals' Fourth Amendment rights and that the presence of apparent contraband in the plain view of the officers was adequate probable cause for a search warrant.

## Fire Inspections

Fire inspections became a function of code enforcement in July 1991, when North Carolina adopted the Southern Building Code, and in April 1992, when a fire prevention safety code volume (Volume V) was first added to North Carolina's building code. On January 1, 2002, when North Carolina adopted the International Building Code, it included a volume dealing with fire safety issues.

The legal basis for code enforcement begins with three powers set forth in the constitutions of the United States and North Carolina: the power to *tax and spend*, the power of *eminent domain*, and the power to *police*. The tax-and-spend power is the authority of a governmental unit to tax its residents in order to provide funding for governmental functions, such as code enforcement. The power of eminent domain is the authority granted to governmental units to condemn and take property in order to further some legitimate governmental purpose as long as the owner of the condemned property is adequately compensated for its value. In code enforcement, this power typically involves removing something which is a nuisance or a danger (see, for example, Section 203.1.1.1.2, North Carolina Administrative Code and Policies; NCGS 58-79-20). The third power, the one at the heart of code enforcement, is the police power. This is the constitutionally granted authority to regulate behavior.

The police power is subject to additional constitutional controls. Two of them are used to prevent abuse of this power: due process of law and equal protection under the law. Due process is part of the Fifth Amendment to the Constitution of the United States and is also granted in North Carolina's constitution. Equal protection under the law was established in the Fourteenth Amendment to the Constitution of the United States and is likewise guaranteed in North Carolina's constitution.

When courts examine a set of facts to determine whether or not the set meets a particular legal requirement and there is no readily available answer to the question, the courts develop tests, or lists of specific questions, which they can use to analyze sets of facts in order to arrive at answers. These tests are ordinarily identified by the number of questions in the test. Due process issues arise in an almost infinite variety, so the courts have subdivided due process into two categories: Substantive Due Process and Procedural Due Process. Substantive Due Process ordinarily involves an attack on the wording of a law, rule or regulation and the courts use a four-pronged, or four-questioned, test when analyzing such a disagreement issue. In a code enforcement/ inspection context, such a dispute ordinarily arises from the wording of a provision of a code. In such an instance, the court asks the following four questions in no particular order. Does the regulation bear a substantial relationship to the police power objective? In other words, does the regulation attempt to control construction activities or does it attempt to control something else? Is the extent of the regulation reasonable? Or are there some limits to its extent, and is it restricted to the proper subject matter? Is the regulation sufficiently specific to be understood by both sides in the controversy (for example, by the CEO and the workman)? Lengthy and complex regulations are particularly vulnerable to attack because of misunderstanding by one or both parties.

A fourth question is whether or not the regulation is being enforced in an arbitrary or capricious manner. In other words, is everyone required to do or not do this, or are some people or groups exempted from abiding by the regulation? If the court can answer "Yes" to all four questions, then due process exists. Only a single "No" is necessary for a ruling that the regulation in question violates due process. Procedural due process involves the managing of the dispute and the courts use a three-prong test. The first question is was the accused informed of the charges/accusations against him? The second question is was he afforded an opportunity to present his side of the dispute? The third question is was he allowed to confront the witnesses against him? All three questions must be answered in the affirmative by the court in order to pass this due process test.

Equal protection under the law is suggested in other portions of the Constitution of the United States, but until 1868 it was not set out specifically as a right entitled to constitutional protection. In 1868, the Fourteenth Amendment was ratified in order to assure that the constitutional rights of persons freed from involuntary servitude subsequent to the War Between the States were protected and guaranteed. Simply put, the Fourteenth Amendment says, among other things, that the Constitution applies to all residents of the United States all of the time.

Two intensely litigated questions have arisen under the Equal Protection clause of the Fourteenth Amendment and have had an influence upon fire service operations. First: When is it acceptable to treat one person or group of persons differently from the majority? And second: Can the various branches of federal and state government trade responsibilities? At present, the most notable example of the first question involves the issue of affirmative action. The courts are currently debating the question, When does the need to help one group of people without helping another cross into unconstitutional discrimination? The second question has been addressed in North Carolina in the code enforcement context. The creation of law is a legislative function, as directed by the United States and North Carolina constitutions. Remember that both constitutions forbid the intermingling of the three branches of government: legislative (Congress/General Assembly), executive (president/governor), and judicial (the court system). This concept is called the separation of powers. In North Carolina, the legislature handed its constitutional responsibility for creating the building code to a portion of the executive branch of state government (the Department of Insurance). Legal assaults were launched by those opposed to a statewide building code, asserting that the code itself was unconstitutional because it had been created by the executive branch of government, not the legislative one. The courts, however, upheld the actions of state government and said that practical necessity required such a course of action. A statewide building code was clearly in the best interests of the citizens and residents of North Carolina, and the only practical way to create one was to deliver the responsibility for doing so to the experts—in this case, the Department of Insurance. This transmission of authority by legislative action is known as "enabling legislation." However, once the code was written and approved at DOI, it still had to be passed back through the General Assembly for approval. This process preserves the constitutional requirements of separation of powers by making the final approval of the code a legislative activity.

Below the courts, the next level in the legal structure for code enforcement (fire inspections) is the General Statutes. In Article 9 of Chapter 143, the legislature established the Building Code Council (NCGS 143-136, 137) and authorized it to write the Building Code (NCGS 143-138). In NCGS 143-137(c) and (d), the responsibilities for technical support and finances for the council are given to the Department of Insurance (DOI). The Council, together with the Engineering and Codes Division of DOI, write the code. All of these organizations are currently located within the Office of the State Fire Marshal (OSFM) within DOI. NCGS 143-138 authorizes criminal penalties for code violations—a $50.00 fine for each violation which lasts up to thirty days. Each subsequent thirty-day violation is considered a separate criminal misdemeanor and is also subject to a $50.00 fine. Also included in NCGS 143-138 are exceptions to coverage under the code, some high-rise life safety requirements, and general descriptions of subjects to be included in the code.

NCGS 143-151.8 (together with statutes following) establishes the North Carolina Code Officials Qualification Board. Located within OSFM, it is responsible for establishing the standards for licensing CEOs (including fire inspectors), conducting the necessary examinations, and taking disciplinary actions against CEOs who violate the rules. The operating rules for the board are found in Title 11, Chapter 8, NCAC, Sections .0500 through .0800. Found also in this chapter are the application and qualification requirements for each trade and level of CEO, including Fire Inspectors). North Carolina currently recognizes three levels of certification in each inspection trade, including a fire inspector: Level I the entry level, Level II the intermediate level, and Level III the highest level. Inspectors certified for Levels I and II are restricted with regard to the sizes and types of building which they can inspect; a Level III inspector is authorized to inspect any structure subject to the code. Title 11, Chapter 8, NCAC also explains in some detail the inspection authority of each level of certification.

Below the enabling legislation is the Building Code itself. Once written and approved by the Building Code Council (with the assistance of the DOI), the code is published for public comment and then submitted to the legislature for approval (ratification). When the code, which was created by the executive branch of state government, is returned to the legislative branch for final approval, the constitutional requirement of separation of powers between legislative, executive, and judicial branches of government is reestablished. While the code was created to deal with building issues, the legislature has enacted some statutes with direct impact upon code enforcement. NCGS 143-135.1 exempts state-owned buildings from inspection by local CEOs; NCGS 143-139 provides for the enforcement of the code; and NCGS 115C-525 sets the rules for fire safety inspections for schools (two inspections during the school year, at least 120 days apart). Other regulations are found in the North Carolina Administration and Enforcement Requirements Code, which is a volume of the North Carolina State Building Code and the North Carolina Administrative Code and Policies volume.

While the Building Code is intended to deal with construction issues on a statewide basis, North Carolina's geographical diversity has made local modifications to the code necessary. These local modifications appear occasionally in the code itself, but more often they appear as legislation known as "local acts." Because the local acts are not applicable statewide, they exist as the next-lower level on the ladder of inspection legal authority. Upon beginning work, the prudent inspector should always consult his or her inspections department regarding the presence local modifications to the code. Individual counties and municipalities occasionally enact their own additions to the code as well. These additions or modifications must meet or exceed existing statewide code standards in order to be enforceable. Standards may be tightened at the local level, but they cannot be loosened without the approval of state government. Many counties and municipalities, for instance, have created local ordinances which stiffen the penalties for code violations and add requirements for employment above those specified in Chapter 143. This is especially true in areas of the state with high population densities or heavy industrialization.

North Carolina law does not prescribe an organizational structure for an inspection department. In NCGS 160A-411 and NCGS 153A-351, municipalities and counties, respectively, are afforded considerable latitude in the ways they organize their departments. As long as certain functions are fulfilled, the organization of the department is left pretty much to the discretion of the AHJ. The department must manage the permitting process, conduct the inspection, and take appropriate enforcement actions to approve the work or to correct discrepancies. At the local level, the fire inspector, if a member of a fire department which is part of the AHJ, can wear two hats, reporting to the office in charge of inspections as well as the fire chief. The prudent inspector should inquire ahead of time how the chain of command is structured under these circumstances. When inspecting, the behavior of the fire inspector is subject to the rules of the Qboard and the Council and the information gathered pursuant to the inspection activities should be furnished to the inspection department having jurisdiction, and if requested, to the fire department. While the fire chief possesses overall authority over the inspector, his/her authority to inspect is derived from the AHJ through its inspection department. It should be remembered that fire inspections are code enforcement activities and that the inspectors are trained and certified to perform code enforcement activities and their enforcement activities overseen by the Qboard and not by the fire department or the Fire and Rescue Commission. That being said, the organizational flexibility allowed to North Carolina's inspection departments in theory might not preclude an arrangement by which the department's responsibilities were bifurcated such that a fire department officer holding a higher level of certification (ideally Level III) could be designated by the AHJ as the fire inspections supervisor and custodian of the AHJ's fire inspection records.

A new and emerging area of code enforcement is a minimal-standards housing code. Such codes are currently local matters, enacted by AHJs as local ordinances. Because most code enforcement, with the exception of commercial occupancies, ends when the project is completed, there has not existed a mechanism to insure that certain types of residential dwellings are maintained by their owners in a code-compliant condition once construction has been completed and the property occupied. Until recently, the only remedies available to tenants occupying substandard housing were to leave and risk a fight with the lessor (landlord or landlady) over the unfinished term of the lease, file a lawsuit, stage a rent strike, or complain to local code authorities and have the place condemned. None of these solutions have really worked, because they either cost the tenant his or her home or they cost too much money to pursue. A minimal-standards housing code provides a mechanism to oversee residential lessors and to assure tenants, especially low-income ones, of adequate housing without automatically placing the tenant in an adversarial relationship with the lessor. Because of ongoing problems with fire losses involving high-density residential occupancies, fire inspectors should begin planning for the time when fire code inspections of such occupancies become part of the ordinary inspection rotation of the AHJ.

An important, but frequently ignored, aspect of fire code enforcement is the concept of fire limits. These are areas within a municipality where special attention must be given to wood frame construction. NCGS 160A-435 requires that every incorporated municipality enact ordinances describing its primary fire limits, which are the principal business areas of the municipality. Within the primary fire limits, no frame or wooden building may be erected, altered, repaired, or moved without a permit approved by the local inspection department, its AHJ, and a designee of the DOI. The governing body of the AHJ is granted authority to enact any additional regulations deemed appropriate to manage these matters (NCGS 160A-436). The municipality also may designate secondary fire limits (NCGS 160A-437) and impose whatever regulations may be deemed appropriate to control the erection, alteration, repair, or movements of frame and wooden structures within this area. If a municipality refuses or neglects to establish its primary fire limits after having been so notified in writing by the DOI, the Commissioner of Insurance may designate those limits if doing so is deemed in the best interest of the residents of the AHJ. Section 203.1.3, NC Administrative Code and Policies requires that, prior to approval, all permits relating to wood frame buildings located within primary fire limits be reviewed by the inspections department with jurisdiction, the governing body of the AHJ, and the Commissioner of Insurance (or a designee).

Because of the long-standing historical relationship between major fires and code enforcement, it is becoming increasingly apparent that code enforcement and C&O investigations have common interests. History has demonstrated that the results of C&O investigations have often led to the creation of or significant modifications to building codes. North Carolina appears to be recognizing this seemingly symbiotic relationship by locating code enforcement offices within OSFM and delivering disciplinary authority over both CEOs and fire investigators to OSFM, as specified in Section 203.1.1.1, NC Administrative Code and Policies.

## Chapter 6

# Apparatus Operations

## Introduction

This chapter will examine the North Carolina laws relating to the operation of fire department vehicles, including rules of the road, licensure of operators and apparatus, traffic direction, and auxiliary equipment installed on a privately owned vehicle (POV). Matters relating to liability will be discussed in a separate chapter.

## Rules of the Road

The operator of a fire apparatus, whether working in an emergency mode or in a routine mode, must obey all rules of the road unless a general statute specifically exempts the operator from following a rule under circumstances then existing. If the exemption is warranted, the action involved in the rule must be performed *without endangering others* on the road at that time. It is vital for the operator of an apparatus to keep in mind that most modern engines, tankers, quints, aerial devices, and support vehicles weigh as much as light and medium tanks of World War II vintage and cannot be stopped nearly as quickly as those armored vehicles. There is ample nationwide evidence in support of the principle that a POV suffers horribly in a collision with a fire apparatus or other emergency vehicle.

NCGS 20-145 allows fire department vehicles to exceed speed limits when responding to fire alarms. However, the statute requires that the responding vehicles be operated with due regard for the safety of others and specifically does not protect the operators from the consequences of the reckless disregard of the safety of others, even when responding in an emergency mode. The statute does not define a "fire alarm," but the courts are recognizing the term in a general sense: it includes any *emergency* to which a fire department may be summoned, whether it is a fire or some other situation endangering life and/or property. However, the courts are also examining the question of what is a proper "summoning" of an emergency responder and it appears that the responder must be alerted and directed to the scene of the emergency utilizing the dispatching system in use by the authority having jurisdiction over the responder, typically a municipal or county-based system. The act of alerting and dispatching the responder is an important component of the "qualification" of the response as a valid one and thus entitling the response to the protection of the statute.

Right-of-way rules upon public streets, roads, and highways have been modified by NCGS 20-156 to accommodate the needs of emergency responders. In NCGS 20-156(b), drivers are required to yield the right of way to an emergency vehicle when the vehicle is giving appropriate warnings by lights and by an audible warning which can be heard from a distance of not less than 1,000 feet ("running with lights and siren"). When appropriate warnings are being given, an emergency responder may proceed through an intersection or place controlled by stop signs or some form of light system. However, the statute *does not* relieve the operator of the emergency vehicle from the duty to operate the vehicle with due regard for the safety of others, nor does it protect the operator from the consequences of any arbitrary exercise of the right-of-way privilege (trying to force the issue of the right-of-way). The courts have emphasized that a motorist is under no obligation to yield to an emergency vehicle unless proper warnings are being given.

NCGS 20-157 is concerned with motorists operating near emergency response vehicles. In subsection (a), motorists are required to move as far to the right-hand side of the road as is safely possible, to stop, and to remain stopped until the emergency vehicle passes their location if the approaching emergency vehicle is giving appropriate visual and audible warnings. This rule does not apply to emergency responses on four-lane, limited-access highways separated by median strips. Subsections (b) and (c) describe the distances at which a driver, other than one on official business, may follow an emergency apparatus or may stop and park (one block in town and 400 feet outside town) in the vicinity of the emergency vehicle. Subsection (d) forbids a motorist from driving over a fire hose or other equipment being used at a fire and from blocking fire apparatus or equipment from a water supply source. Subsection (e) prohibits a motorist from parking within 100 feet of an emergency vehicle when it is being used to investigate an accident or render assistance to victims of the accident. In 2001, subsection (f) was added to the statute. It imposes additional requirements on motorists when approaching emergency vehicles parked in the roadway or within twelve feet of the roadway and which are operating appropriate warning lights. If the road has at least two lanes in the direction of the approaching vehicle, the approaching motorists must move to a lane farthest from the emergency vehicle(s) until safely past them. If the road is a two-lane road, the motorist must slow to a speed that is safe for existing conditions and proceed past the emergency vehicle(s). Both of these rules are subject to any traffic directions being given by the emergency responders. In other words, if the emergency responders are directing traffic at the scene, the motorist must follow the traffic directions of the emergency responders, including stopping all traffic until traffic can pass the scene without endangering emergency responders or anyone else attempting to use the roadway in question. These basic rules are subject to continuing amendment by the General Assembly in an effort to better protect emergency responders and should be monitored closely

Parking issues are discussed in NCGS 20-162. It is unlawful to park within fifteen feet of a fire hydrant or the entrance to a fire station. The statute does give AHJs the authority to shorten the parking distances from fire hydrants by local ordinances, however. A motorist is prohibited from parking in *properly designated* fire lanes. But a motorist may park to load and unload as long as someone attends the parked vehicle. By parking in a fire lane, a motorist has appointed the AHJ as his or her agent for purposes of towing and storing the parked vehicle, and neither the AHJ nor its agents who tow the vehicle can be held accountable for the consequences of the towing unless the towing process amounts to wanton conduct or intentional wrongdoing.

## Traffic Control

NCGS 20-114.1 establishes a framework for traffic control by emergency responders. Subsection (a) requires all persons to comply with the lawful traffic orders of law enforcement officers or other traffic control officers who possess the authority to direct traffic. Subsection (b) adds uniformed members of paid or volunteer fire departments to the group of traffic control officers as long as they direct traffic while performing their duties as firefighters or rescue squad members. This subsection limits their authority as traffic direction officers to their respective emergency responses and does not automatically appoint them full-time traffic control officers. Firefighters and rescue squad members have authority at the scenes of their emergencies and activities *in connection with* their emergency responses. Arguably, they are also authorized to manage traffic at sites remote from the emergency scene (during HAZMAT emergencies, aircraft accidents, and major fires, for example). Apparently, there is no court guidance regarding the definition of a "uniform" within the meaning of the statute. Clearly, a "uniform" is some sort of identifying clothing, however. For firefighters or rescue personnel, a uniform (or turnout gear) could be a vest or headgear which distinguishes the member from bystanders, or some other clothing which distinguishes the person engaged in the traffic control from the bystanders and indicates membership in one of the responding units.

A thorny question which occasionally arises in traffic control concerns command: Who is in charge? There is little statutory guidance outside HAZMAT situations. Often the issues of traffic control command are worked out as a result of time and experience within the AHJ. Ordinarily, if the event is a major one, the agency with the most assets on the scene (usually the fire department) assumes command over all aspects of the incident and then delegates responsibilities. Typically, in situations where the responding fire department is a paid one, the law enforcement agency of the AHJ controls traffic while the fire department performs its traditional duties associated with the emergency response. The answers to command issues become less clear when the fire department in charge is a volunteer unit. This is the situation where experience (and possibly preplanning) can develop guidelines to facilitate effective traffic control.

Another problem which sometimes arises is a situation in which the emergency responder (typically a VFD) is asked by another agency (usually law enforcement) to remain at a scene and continue to exercise traffic control authority after the emergency response is completed. Usually, this is because some traffic management system (lights, signage, railroad safety equipment, for example) has been damaged. Arguably, under the statute, the authority of the emergency responder to direct traffic ends when the emergency response ends. Additionally, common sense says that a fire department, for instance, may not be adequately protecting is jurisdiction by leaving resources (personnel and equipment)to watch over broken equipment. After all, traffic control is one of law enforcement's primary duties. LEO's are trained and certified to do it. As a practical matter, a prudent incident commander should notify the dispatching authority as soon as the response activities are concluded and "clear the scene" promptly in order to avoid becoming involved in unauthorized traffic control situations.

# Operator Licenses

In order to operate a motor vehicle upon the highways of North Carolina, the operator must possess a valid operator's license (NCGS 20-7). The statute establishes various types of licenses. The three basic licenses are known as Class A, Class B, and Class C. A Class A license entitles its holder to operate a Class A vehicle which is exempt from commercial driver's license requirements, a Class A vehicle which has a combined gross vehicle weight rating (GVWR) of less than 26,001 pounds and which includes as part of the GVWR a towed unit of at least 10,001 pounds GVWR, and any Class B or Class C vehicle. Class B licensees may operate any Class B vehicle which is exempt from commercial driver's license requirements and any class C vehicle. In addition, any Class C operator who is a volunteer member of a fire department, rescue squad, or EMS provider and who is in the performance of duties may operate any Class A or Class B vehicles of the department or any combination of Class A and B vehicles of the department (NCGS 20-7(a)(3), 20-37.16(e). Any new resident of North Carolina licensed in another state must obtain a valid North Carolina license within sixty days after becoming a resident. Arguably, a member of the Armed Forces of the United States on active duty in North Carolina, who is serving as a member of a volunteer fire department, rescue squad, or EMS provider, who has a valid driver's license from another state, and who is also a resident of another state can operate the same vehicles which a member of a volunteer fire department, rescue squad, or EMS provider who possesses a valid North Carolina Class C license is authorized to operate.

NCGS 20-4.01 includes definitions of types of motor vehicles. A Class A vehicle is either a combination of motor vehicles which has a combined GVWR at least 26,001 pounds and which includes as part of the combination a towed unit of at least 10,001 pounds GVWR, or a vehicle which has a combined GVWR of less than 26,001 pounds and which includes as part of the combination a towed unit of at least 10,001 pounds GVWR. A Class B vehicle is a single vehicle with a GVWR of at least 26,001 pounds or a combined vehicle with a towing unit of at least 26,001 pounds GVWR and a towed unit of less than 10,001 pounds GVWR. A Class C vehicle is any single vehicle not included in Class B or a combination of motor vehicles not included in Classes A or B.

One other license sometimes utilized in North Carolina's fire departments is the Commercial Driver's License (CDL). The license authorizes the holder to operate commercial vehicles. A commercial vehicle is defined as one which transports passengers or property and which:

a. has a combined GVWR of at least 26,001 pounds, including a towed vehicle of at least 10,001 pounds GVWR; or
b. is a Class B vehicle; or
c. is a Class C vehicle which transports either at least sixteen passengers (including the driver) or HAZMAT and which is required to be placarded.

Frequently, these licenses are required by paid departments which possess large, articulated aerial devices or anticipate acquiring such an apparatus. Otherwise, it appears that a Class A or B license is sufficient. Volunteers, of course, need possess only Class C licenses.

## Registration

Cities, towns, counties, incorporated emergency rescue squads, and rural fire departments, agencies, or associations are entitled to permanent registration plates for their vehicles. Such a plate may be obtained by providing proof of ownership and of financial responsibility and by paying a fee. The plate may be transferred to a replacement vehicle owned by the registrant if the vehicle is of the same classification as its predecessor (NCGS 20-84). The law regarding registrations and privately incorporated EMS providers appears unclear. Such providers are not included in the list of authorized recipients of permanent plates, unless they are included as "incorporated emergency rescue squad(s)."

## Privately Owned Vehicles (POVs)

Horns, sirens, or similar audible warning devices are authorized for use on POVs owned by chief officers of fire departments and rescue squads, fire marshals, and emergency management coordinators when they are performing their official or semiofficial duties. Two questions unanswered by this statute (NCGS 20-125) concern its application to EMS chief officers and the definition of "semiofficial duties." Arguably, EMS can be included in the definition of "rescue squad," and "semiofficial duties" can include such activities as parades or fund-raising functions.

Electronically modulated headlamps ("wigwags") are not authorized for use on POVs in North Carolina. NCGS 20-130(d) authorizes them for use upon motorcycles, law enforcement vehicles, fire department vehicles, rescue squad vehicles, ambulances, and vehicles operated by fire marshals and emergency management coordinators. The equipment itself must be of a type approved by the commissioner of motor vehicles. The installation of these devices on POVs operated by members of paid or volunteer fire departments constitutes unlawful modification of federally mandated safety equipment. Members of paid or volunteer departments, *while in the performance of their duties*, are authorized to use red lights on their POV's.

Currently, a popular light device operated by fire service personnel on POVs is a set of flashing strobe lights, either red or white and mounted in the front (and occasionally the front and rear) of the vehicle. NCGS 20-130.3 prohibits the display of white or clear lights on the rear of a vehicle when it is being operated in a forward direction upon the highways of North Carolina—with the exception of the lights used to illuminate the license plate. Arguably, this statute prohibits the use of clear or white strobe lights on the rear of a vehicle while it is moving forward.

North Carolina's courts have been quite clear regarding POVs owned and operated by fire service personnel when responding to alarms. Their drivers must follow all rules of the road, and the vehicles themselves are not entitled to any of the privileges afforded emergency service vehicles. For example, the display of a red light does not convert a POV to an emergency vehicle, regardless of the identity of the operator. Responses by POVs should be an important consideration in the SOGs of every fire department in North Carolina.

## All-Terrain Vehicles

NCGS 20-171.23(b) allows fire, rescue, and EMS department members to operate ATVs owned or leased by their departments, or ATVs operated under the direct control of a department on a public highway where the speed limit is 35 MPH or less and on a non-limited access highway in order to move the ATV to a highway where the speed limit is 35 MPH or less. All the ordinarily applicable rules and laws relating to the operation of ATV's remain in effect. The ATV must have operable front and rear lights and a horn and must be operated in accordance with the manufacturer's recommended performance requirements. The operator SHALL possess a badge or departmental identification card when operating the ATV under these circumstances.

# Chapter 7

# Mutual Aid

"Mutual aid" is a term with many "official" definitions. But in the context of emergency services it simply means the support which two or more agencies agree to provide each other upon request. In today's world, when the demands placed upon the fire service are increasing exponentially and available assets are not, mutual support among departments has become a necessity in many jurisdictions. Mutual support can occur among paid departments, volunteer departments, and other emergency service agencies.

North Carolina recognized both the need for and the effectiveness of support arrangements and by 1939 had enacted the predecessor to Article 80 of Chapter 58 of the General Statutes. The article establishes and provides rules for the State Volunteer Fire Department (SVFD). The stated purpose of the SVFD is to provide fire protection for unincorporated areas of the state and provide additional assistance to any area within the state, whether a municipality or a county, in case of emergency (NCGS 58-80-1). This statute appoints the Commissioner of Insurance as the State Fire Marshal. The SVFD consists of all personnel serving in organized fire departments who are also members of the North Carolina State Firefighter's Association (NCSFA) and whose municipalities and counties have, by resolution of their governing bodies, subscribed to and endorsed the provisions of Article 80 of Chapter 58. Such an endorsement is not compulsory (NCGS 58-80-15). A municipality may withdraw from the SVFD voluntarily by resolution of its governing body and by serving notice of its withdrawal on the State Fire Marshal. The withdrawal is effective sixty days after its receipt by the State Fire Marshal (NCGS 58-80-20).

The command structure consists of a chief of the SVFD who is the State Fire Marshal, with municipal chiefs serving as assistant chiefs of the SVFD, municipal assistant chiefs serving as deputy chiefs of the SVFD, and other officers and firefighters in their respective departments serving with their customary titles and duties. The statute even establishes a chain of command for initial responses. The ranking officer of the first-in department has command until the arrival of higher-ranking officers. The statute designates the chief present at an emergency as the officer in complete charge of all operations (NCGS 58-80-10).

The subscribing municipalities retain full control of the decision to send assets outside their jurisdictions and must designate three people (a primary person and two alternates) to authorize the dispatch of assets from the municipality. The municipality must furnish this information to the commissioner of insurance and the NCSFA, whose secretary is to furnish it to all counties and municipalities which have endorsed Article 80 of Chapter 58.

The SVFD has no authority to provide assistance within a county which has not accepted the terms of Article 80 of Chapter 58. However, the statute does not prohibit a municipality from providing assistance on its own initiative (NCGS 58-80-30).

A county may elect to accept the benefits of the SVFD by resolution of its commissioners and may enter various agreements and/or contracts regarding payment for services rendered in accordance with them (NCGS 58-80-35). The statute authorizes the county to contract with any SVFD member municipality to provide services under financial terms that they are able to negotiate between themselves. If the emergency to which the county summons the department occurs within a municipality within the county, the municipality requesting the help must also pay the costs.

It is absolutely vital that a municipality responding to an emergency in another municipality not leave itself unprotected. NCGS 58-80-40 prohibits a municipal department from assisting another municipality more than two miles from its municipal corporate limits without leaving behind sufficient assets to adequately protect its own municipality. The decision regarding the sufficiency of the assets left behind is made by the chief of the responding department. Members of the SVFD who respond in a mutual aid context, whether paid or volunteer municipal firefighters, carry the same rights, authority, and immunities they would carry if they were responding within their home jurisdictions. Their activities are deemed the exercise of a governmental function, affording them the privileges and immunities associated with performing governmental functions.

Members of the SVFD who are injured while responding in a mutual aid context are covered by the Workers' Compensation Act, and the municipality for which the injured firefighter is responding is liable for the compensation (NCGS 58-80-50). In order to fund participation in the SVFD, counties and municipalities are authorized to appropriate funds from property taxes and from other revenue sources unless otherwise prohibited by law (NCGS 58-80-55). The governor is authorized to provide the NCSFA at least partial compensation for the expenses it incurs while fulfilling its role in the SVFD (NCGS 58-80-60).

In 1965, the General Assembly revisited mutual aid with NCGS 58-83-1. The statute authorizes counties, municipalities, fire protection districts, sanitary districts, or incorporated fire departments to send *or decline to send* assets beyond the territorial limits of the area within which they normally operate. While operating outside its normal response area, any such department retains all authority, rights, privileges, and immunities it would otherwise hold, including Workers' Compensation coverage. The statute likewise allows the political subdivision whose department is responding to retain all of its ordinary rights, authority, privileges, and immunities.

Notwithstanding statutory rules governing mutual aid, it is vital that all such arrangements be set down in writing—in the form of a contract. By creating a statute with general rules, the General Assembly has wisely left the departments with great flexibility regarding the terms and conditions they can incorporate into their mutual aid agreements. Basic North Carolina contract law requires that all parties in a contract agree to the terms of the contract and that consideration (something of value) be exchanged between the parties in order for a valid contract to exist. In a mutual-aid contract, the most common type of consideration is the promise of support by each party in the contract to the other parties in the contract. In some instances, things of value other than mutual promises can be exchanged by the parties, typically funds or equipment. All mutual-aid agreements should be reviewed carefully by the parties and, if necessary, by their respective attorneys before being adopted. The governing body of each party in the agreement should also review the proposed agreement and vote its approval before a representative of the department signs the agreement and thereby obligating the department to participate in it.

A growing trend in North Carolina is the countywide mutual-aid agreement, usually originating somewhere within county government. Properly organized and drafted, these agreements can result in well-organized countywide responses, with "automatic" mutual-aid responses guaranteed between departments under certain conditions. Such an agreement also provides a single structure for dispatching agencies to organize efficient and rapid mutual-aid responses. However, sometimes the governmental body which produces the agreement will add conditions to it which may not be in the best interests of all parties. Therefore, each party should examine any proposed agreement carefully to be sure it does not contain any objectionable terms. Problems that could arise from such terms should be negotiated by the participants prior to their acceptance of the agreement. Once an agreement with objectionable terms has been accepted, it becomes very difficult to alter it.

## Chapter 8

# Nonprofit Corporation Operations

All North Carolina VFDs must be nonprofit corporations in accordance with North Carolina law. The basic rules for creating and managing nonprofit corporations are found in Chapter 55A of the North Carolina General Statutes, which is one of the chapters of the General Statutes which utilizes a three-number system of designation instead of the more traditional two-number system. In North Carolina, most corporate administrative activities, including the creation of the corporation itself, are conducted under the overall supervision of the Business Entities Division of the Office of the Secretary of State.

## Creation

To create a corporation, documents called Articles of Incorporation are filed with the Secretary of State (NCGS 55A-2-02). The statute contains a list of the information required on the forms, together with the required filing fees (NCGS 55A-1-22). There is no requirement that an attorney be involved in the filing of the articles, but frequently an attorney's involvement at the beginning can save time and cost at a later date if an avoidable mistake requires correction. Once the articles have been filed with the Secretary of State, a copy of them, certified by the Secretary of State, is returned to the corporation, thereby confirming its existence. Now the corporation is authorized to engage in the activities for which it was created.

Under earlier law, a corporation could not engage in its authorized activities until it had recorded its articles in the county where it maintained its registered office. While this is no longer required, a VFD should record its articles in the

Office of the Register of Deeds of the county where it has its main station and primary office in order to assure ready availability, especially if the originals are lost or destroyed.

In North Carolina, a corporation is a free-standing entity, just like any person, and it is therefore required to have both a unique name and an entity inside the state which can receive correspondence or legal documents. Known as the corporation's "registered agent," this entity must be a person or another corporation and must be identified in the articles by name and by both physical and mailing addresses. Frequently with VFDs, the original registered agent listed in the articles is a respected older member of the corporation who has died or has left the department. North Carolina law requires that a "legitimate" registered agent be maintained at all times and that all changes of the registered agent be reported to the Secretary of State whenever they occur (NCGS 55A-5-01). VFDs are notorious for not following this requirement, largely because they forget about it, but it is an essential part of efficient corporate operation. The statute also requires that a corporation maintain a "registered office" in the county of its principal place of business. For a VFD, this should not be a problem: The office is the headquarters station. Current mailing and physical addresses of the registered office are to be furnished to the Secretary of State.

## Membership

NCGS 55A-6-01 (together with the statutes following) sets out the basic rules for membership in nonprofit corporations. Because these corporations are not allowed to issue stock, they are permitted to allow membership in them and to issue certificates indicating membership. The statutes permit a wide variety of memberships within a corporation, and rules regarding membership can be established within the corporation by its Articles of Incorporation or bylaws. The recommended method for handling membership rules is to reference the existence of members in the articles and then designate the classifications of members in the corporate bylaws. By so doing, the corporation can adjust the classifications of its members without undertaking the cumbersome process of amending its articles. Corporate bylaws, the day-to-day operating rules of the corporation, are not required by the General Statutes, but the corporation which does not enact bylaws must operate strictly according to the detailed rules set out in Chapter 55A. A much-preferred procedure is for the corporation to write its own set of bylaws. Certain rules insist Chapter 55A will still apply, but by writing its own bylaws, the VFD will create operating procedures more suited to its needs, purposes, and duties. Article 10 of Chapter 55A sets out some basic rules for amending articles and bylaws and should be examined by the VFD before enacting any amendments.

Members can be removed from nonprofit corporations (VFD's) as long as the removal/expulsion is fair, reasonable, and in good faith (NCGS 55A-6-31). Most serious accusations against VFD's regarding removal from membership center around an argument that the removed member was denied due process of law in the removal process. Due process under these circumstances should consist of notification to the accused of the charges, an opportunity to present his/her side of the question, and to do so in the presence of the accusers/witnesses. All this should be set out in some format in the bylaws of the VFD and followed for any such disciplinary expulsions. The existing court cases support this idea.

## Rules, Regulations, and SOGs

It is tempting for a VFD to intermingle its corporate bylaws, rules, regulations, and SOGs. This is *not recommended.* As explained earlier, a VFD is really two organizations operating under a single roof: the corporation, which serves as an overall management agency; and the fire department itself, which delivers the services. Unlike most other nonprofit corporations in North Carolina, VFDs are engaged in activities dealing with some of the most hazardous threats to the life and property of the public. These activities are subject to constant study and, upon occasion, outside regulation. The constant evolution and change in duties, training, and regulations and the fact that a fire department is, in many respects, a paramilitary organization require that the service-delivery component of the organization be flexible in its approach to the performance of its duties and its required training. If the rules, regulations, and SOGs are incorporated into the bylaws of the corporation, any changes to them must be accomplished in accordance with the cumbersome requirements associated with bylaw and article amendment. The chief and his or her subordinate officers should be allowed ample flexibility to adapt the delivery of the services to changing circumstances.

One method of accomplishing this is to make reference to the fire department in the corporate bylaws and then authorize the department's officers to create and maintain such rules, regulations, and SOGs which they deem appropriate and which are subject, ultimately, to the authority of the corporation. These ideas can be discussed and approved by the firefighters themselves at regular department meetings.

The articles and bylaws of the corporation should deal with management and policy issues, and the rules, regulations, and SOGs should deal with the delivery of the services.

# Chain of Command

Every corporation must have a board of directors, except under certain very limited circumstances (NCGS 55A-8-01). The purpose of a board is to set corporate policy and serve as the supreme governing entity within the corporation. The qualifications, terms of office, and method of selection for members of the board are ordinarily explained by the articles and/or bylaws of the corporation. If they are not specified, NCGS 55A-8-02, 03, 04, 05, and 06 will establish these rules. In most VFDs, directors are elected by the members of the corporation. Directors may resign at any time by communicating their resignations to the board, its presiding officer, or the corporation (NCGS 55A-8-07). The presiding officer of the board, and its senior member, is the chairman. The board, either by election or appointment, may designate other officers (for example, vice-chair, secretary, and so forth) or committees, as may be needed. Removal of directors from office can be addressed in the articles or bylaws of the corporation, but if they are not, the procedures in NCGS 55A-8-08, 09, or 10 must be followed.

Next in the chain of command are the corporate officers. Their responsibility is to implement the policies stated and approved by the board. A typical list of officers consists of a president, a vice president, a secretary, and a treasurer, with assistants to each officer that may be needed by the corporation (NCGS 55A-8-40, 41). The officers may be elected or appointed, depending upon the needs of the corporation. The processes of election or appointment are ordinarily addressed in the articles or bylaws. If not explained in the articles or bylaws, the removal of corporate officers must be conducted in accordance with NCGS 55A-8-3.

Members in a nonprofit corporation occupy a unique status. They can be the supreme governing body of the corporation, or they can have little or no authority. Their place in the chain of command is ordinarily determined by the articles or the bylaws. Many VFDs grant much of the policy-making authority to the board, with the board's decisions or recommendations subject to final approval by a vote of the members. The classification, qualification, and selection of members are matters for each corporation/department to determine in its articles and bylaws.

Because VFDs comprise a unique category of nonprofit corporations, they require an additional chain of command—one for the delivery of services. This, of course, consists of the fire chief and the subordinate officers (department officers). In most small VFDs, many of the department officers serve also as corporate officers and members of the board of directors. While this may be necessary for practical reasons, every effort should be made to protect the board, the corporate officers, and the department officers from problems associated with conflicts of interest (NCGS 55A-8-30, 31, 42). VFDs can accomplish this by dividing the duties among as many different people as possible.

# Meetings

Chapter 55A requires that members and directors of nonprofit corporations meet at least once per business year. These meetings can be scheduled as far in advance as the corporation chooses as long as the notice requirements of NCGS 55A-8-22 and 55A-7-01 (together with the statutes following) are met. Most nonprofit corporations hold meetings on a regular basis, typically according to a prearranged schedule (monthly, quarterly, and so forth). As long as the meetings are scheduled and the members and directors are made aware of the schedule, little additional notice is required. Special meetings require notice to all members of the group for which the meeting has been requested. Notice requirements are specified in the articles or bylaws of the organization. If no notice requirements are specified therein, the requirements specified in Chapter 55A must be followed. If the notice requirements are not met, actions taken at corporate meetings can be void or voidable.

Directors may waive notices of meetings and may take actions by the consent of all the other directors without conducting a formal meeting (NCGS 55A-8-23 and 55A-8-21). Such actions require the written consent of *all* directors and the actions taken must be put into writing. If directors cannot attend meetings in person, they may use methods of communication (such as telephone-conferencing) which allow them to communicate with the other directors as if they were present. Whether or not a corporation's meetings are subject to the Public Records Law and Open Meetings Law depends upon the nature of the corporation and purpose for which it was created (NCGS 55A-3-07). It appears that a VFD is not subject to these laws unless it is part of a municipal or is a county fire department (actually part of a political subdivision of the state). As a private nonprofit corporation, its meetings and records would not ordinarily be open to public inspection. Increasingly, however, municipalities and counties which contract with VFDs for services are requiring in their contracts that the VFDs open their meetings and certain financial records to inspection. It is safe to assume that this trend will continue until such requirements are specifically defined and explained by the courts or by the General Assembly. A VFD wishing to protect the confidentiality of its meetings should consult legal counsel when addressing such issues in its contract(s) for services, bylaws, SOGs, rules, or regulations.

## Conduct

Directors and officers of corporations are required by statute to follow certain basic standards of conduct. These standards may be altered by the articles or bylaws as long as the standards set by the statutes are not reduced. Directors are expected to act in good faith, with the care of an ordinary and prudent person, and in a manner which the director believes in good faith to be in the best interests of the corporation. In making decisions the director may rely upon officers and employees of the corporation, legal counsel, accountants, other specialists, and committees of the board if the director believes those persons or groups possess the knowledge and skills needed to give good advice (NCGS 55A-8-30). Conflicts of interest are forbidden, and NCGS 55A-8-31 and NCGS 55A-8-32 explain such rules in some detail.

Corporate officers are also subject to basic standards of conduct explained in the General Statutes (NCGS 55A-8-42), which may be modified by the articles or bylaws as long as the modifications do not reduce the standards set by the statute. An officer must act in good faith, with the care of an ordinary and prudent person in a similar situation, and in a manner which the officer believes in good faith to be in the best interests of the corporation. In making decisions, the officer is authorized to rely upon input from other officers or employees of the corporation and from legal counsel or other specialists as long as the officer believes that the information furnished about a matter is in the best interests of the corporation. Neither a director nor an officer may rely on such input if the director or officer has knowledge of the matter which may call the input into question. A decision made under these circumstances may expose the director or officer to personal liability for its consequences. The obvious lesson is that officers and directors should be honest, candid, and open in their discussions.

NCGS 55A-8-60 allows for certain types of immunity from liability for officers and directors of nonprofit corporations under the conditions set forth in the statute.

## Assets

Like all corporations in North Carolina (both for-profit and nonprofit), VFDs are allowed to acquire, encumber, and dispose of their property (assets) (NCGS 55A-12-01). However, if the VFD is disposing of property in circumstances other than those involving the regular course of its business, the disposal of the property must be conducted in accordance with NCGS 55A-12-02. The statute describes the requirements for approval of the disposition. If a vote of the membership is required, special notice of the meeting to vote must be sent to the members, in accordance with NCGS 55A-7-05. If the intent is to dispose of all the assets of the VFD and the disposal is not to be conducted within the ordinary course of its business, notice must be sent to the Office of the Attorney General for review of the transaction prior to its implementation (NCGS 55A-12-02).

## Corporate Dissolution

Just as North Carolina law provides for the creation of corporations, it also provides for their "uncreation," or dissolution, in Article 14 of Chapter 55A. In abbreviated form, the process consists first of making the decision to dissolve the corporation, typically by vote of the members and directors. Once the decision has been made, a plan of dissolution must be developed and approved. This plan provides a mechanism for collecting the assets of the corporation, for paying all outstanding debts, and then for distributing the remaining assets in accordance with the articles and bylaws of the corporation. During this process, a document called the Articles of Dissolution must be filed with the Office of the Secretary of State. The articles must contain, among other things, the plan of dissolution. When the articles are approved by the Secretary of State, the corporation is deemed dissolved, although the completion of the corporation's business may continue after the Secretary of State's approval. Once the outstanding bills of the corporation have been paid, the remaining assets of the corporation may be distributed. They may not be distributed to members or former members of the corporation unless the recipient is the United States of America, a state, a charitable or religious corporation (for example, another emergency service provider), or a person who is exempt from taxation pursuant to Section 501(c)(3) of the Internal Revenue Code of 1986 or any successor section thereof.

## Distributions and Payments

Distributions and payments by nonprofit corporations are controlled by Article 13 of Chapter 55A. A nonprofit corporation may pay reasonable amounts to members, officers, and directors for services rendered or for other value received (for example, land or equipment) and may reimburse its members for expenses incurred in the course of conducting the business of the corporation. The corporation may also make distributions to entities which are tax-exempt under Section 501(c)(3) of the Internal Revenue Code or which are organized for one or more of the purposes described in Section 501(c)(3) and which will distribute its remaining assets upon dissolution in accordance with Section 501(c)(3). Political subdivisions with which VFDs contract for services are more often placing clauses in their contracts which to restrict the ways an emergency service provider can distribute its assets upon dissolution. All emergency service providers who contract with political subdivisions of the state should be aware of this trend and plan accordingly. As private nonprofit corporations, emergency service providers should be able to dispose of their assets in accordance with their plans for dissolution, even though this may conflict with the terms of the contract for services. This emerging issue has yet to be addressed by the General Assembly or the courts.

## Administrative Dissolution

Nonprofit corporations which refuse or neglect to play by the rules may be dissolved administratively (involuntarily) by the Secretary of State. The grounds for administrative dissolution are as follows:

a. The corporation fails to pay within sixty days any monies due under Chapter 55A; or
b. the corporation lacks a registered agent or registered office within North Carolina for a period of sixty days; or
c. the corporation fails to notify the Secretary of State of a change in its registered office or agent or that the agent has resigned or that the office has closed; or
d. the period of duration of the corporation as stated in its articles has expired; or
e. the corporation knowingly fails to respond to any interrogatories (questions or inquiries) submitted to it by the secretary of state within the prescribed time; or
f. the corporation fails to designate its principal office with the Secretary of State or does not notify the Secretary of State of a change in the location of its principal office within sixty days.

An administrative dissolution restricts the activities of the corporation to those allowed to dissolve the corporation and complete its business. Such a dissolution can be devastating to a VFD because it puts the VFD out of business. The administrative dissolution can be lifted by the offending corpo-ration if it complies with the requirements of the Secretary of State which led to the dissolution in the first place and if it complies with any further requirements made by the secretary of state. All of these activities involve the filing of appropriate documentation with the secretary of state and, possibly, the register of deeds in the county where the corporation has its principal office and registered office.

## Judicial Dissolution (NCGS 55A-14-30, Together with the Statutes Following)

Judicial dissolution occurs when a superior court judge, at the request of the Attorney General, one or more members or directors, a creditor of the corporation, or the corporation itself, issues an order dissolving a corporation or requiring that the corporate dissolution be continued under court supervision. Typically, this kind of dissolution occurs when the corporate leadership is deadlocked, fraud or similar misconduct is occurring (the situation is spinning out of control), or the corporation cannot pay its debts. The court may appoint someone called a *receiver* to manage the dissolution and may give the receiver broad powers to manage it. If the court decides to dissolve the corporation, it will issue a "decree of dissolution," file the decree where appropriate, and direct the receiver or the corporation to proceed with the completion of the corporation's business. When a corporation is in the hands of a receiver, it is said to be "in receivership," which is a condition somewhat similar to a bankruptcy proceeding.

## Merger (NCGS 55A-11-01, Together with the Statutes Following)

Increasingly, VFDs are finding themselves in situations where lack of adequate staffing or financial difficulties are preventing them from functioning effectively. This can lead to loss of an insurance rating with consequent unpleasant consequences for the VFD's constituents. Such a situation leaves the VFD with two options: either go out of business (and abandon its constituents in the hope that another VFD can assume those responsibilities) or merge with another VFD, thereby combining resources and continuing to provide the necessary services. Obviously, a merger is the preferable alternative. A merger is the act of combining two or more corporate entities into a single entity, and because all non-municipal and non-county VFDs in North Carolina must be nonprofit corporations, a merger is the tool typically used to accomplish this task.

Article 11 of Chapter 55A of the General Statutes describes the merger process. According to North Carolina law, this type of activity involves the practice of law and must be performed with the assistance of one or more lawyers. North Carolina law does not allow corporations to "represent themselves" because of the legal nature of corporations. The preparation of merger documentation, the providing of legal advice, and discussion of procedures and legal consequences of mergers is the practice of law. If the merger is "friendly," with no meaningful disputes between the corporations involved, a single lawyer or law firm may be able to handle the entire process. However, should serious disagreements or disputes arise, each side should select its own attorney and the lawyer "in the middle" should step aside until the differences are resolved in order to avoid a conflict of interest.

The first step in any merger is to agree to do it. The VFDs involved should discuss the decision thoroughly and then each board of directors and each group of members entitled to vote must vote upon the decision. If the decision to merge is approved, then the process of merging may begin. What has happened is an agreement in principal to do it and it is now time to bring one or more lawyers into the process if they are not already involved. The Office of the North Carolina Secretary of State oversees this process and can furnish forms, obtainable online, to assist with the process. Because VFDs are not charities, but are charitable nonprofit corporations, the correct forms to use for a merger are found as part of the business entities forms in the Secretary of State's office.

Assuming the merger does not involve disputes, a Plan of Merger (Plan) is the next step (NCGS 55A-11-01). The Plan describes how the merger is to be accomplished. The terms of the Plan must be agreed-upon by all merging corporations. NCGS 55A-11-01 contains a partial list of items to be included in the Plan. Special attention should be paid to personnel matters, corporate board and officer membership, transfer of ownership of land and equipment, pension funds, relief funds, and staffing issues. Because most VFDs are on-behalf-of (o/b/o) agencies for local government, the municipalities and/or counties with which the VFDs have contracts should be advised of the decision to merge if they are not already involved in the process. The Plan should reference OSFM, NCSFA, as well as the municipalities or counties served by the merging departments. The Plan must include a date for the merger to become effective. This is important because the timing of the merger has an effect upon many of the details of the process. The name of the corporation produced by the merger should be selected before the Plan is approved and included in the Plan.

Once the Plan has been approved, Articles of Merger should be prepared for submission to the Secretary of State. NCGS 55A-11-04 contains a list of the information to be included in the Articles of Merger and failure to list these properly can result in the rejection of the Articles of Merger when they arrive in Raleigh. Using the forms provided by the Secretary of State's office can help reduce these types of problems.

Some of the aspects of the Plan can be implemented prior to the filing of the Articles of Merger in order to facilitate the process. However, any such activities should be undertaken carefully and only after consultation with legal counsel. NCGS 55A-11-04 also contains rules for dealing with changes to the Plan which may affect the Articles of merger as well as a situation where a Plan is abandoned and the merger is stopped.

When a merger becomes effective, two entities are now working together as a single entity—the surviving corporation (the corporation which remains in place after the merger) and the merged corporation(s) (the corporations(s) which "disappeared" after the merger). NCGS 55A-11-05 sets out the effects of a merger. The most important thing to understand is that ALL of the assets and liabilities of the merged corporation(s) have become the property of the surviving corporation as of the date of merger. This includes (but is not limited to) relief funds, contents of bank accounts, real property (land and anything attached to the land), contents of stations, equipment (including apparatus), pension fund obligations, training and personnel records, and DEBT. Other things occur, as described in the statute, but understanding how to transfer the assets and debt is an important key to success. If real property is involved in the merger, the surviving corporation should obtain a Certificate of Merger from the Secretary of State and record the Certificate in the Register of Deeds office in each county where any such real property is located. It is very important to understand that a merger will not extinguish any civil or criminal liability for any activities undertaken by or on behalf of any corporation involved in the merger which occurred prior to the merger.

## *Ultra Vires* (NCGS 55A-3-04)

*Ultra vires* is a Latin term used to describe activities by a corporation which are outside its authority. Ordinarily this legal doctrine is used to challenge an act of a corporation when no other means of challenge appears to be available. The statute limits the use of the doctrine to three circumstances:

1. an action by a member or director of a corporation to enjoin (halt or forbid) an act by the corporation;

2. an action by or on behalf of a corporation against an incumbent or former director, employee, officer, or agent of the corporation; and
3. an action by the Attorney General.

The statute authorizes the court to issue appropriate orders to correct the unauthorized corporate action as well as award damages to compensate the corporation for any losses it sustained as a result of the unauthorized corporate action.

## Corporate Records

A thorny question often confronting VFD's and other volunteer emergency services providers is a wholesale request for copies of its corporate records, usually by an AHJ with which it has a contract. One argument which is often made is that a VFD is subject to the Public Records Act and Open Meetings Law. It is not subject to either set of rules. NCGS 55A-3-07 states that only two types of nonprofit corporation are subject to those rules. One is a nonprofit corporation formed to manage funds allocated for economic assistance to a particular industry and the other is one organized at the request of the State of North Carolina to finance projects for public use. AHJ's are now including language in their contracts for services which attempt to require VFD's to follow those rules. What the AHJ's seem to forget or ignore is that much of what is discussed in VFD meetings are matters which may affect the privacy (medical and otherwise) of members of the public, personnel management, and even Homeland Security. Conditions such as the Open Meetings Law and Public Records Act should be watched-for at the time of contract renewal and objected-to strenuously during contract negotiations.

However, members of nonprofit corporations do have certain rights to inspect records. NCGS 55A-16-02 entitles a member to inspect records of the corporation of which he/she is a member at any reasonable time and place and NCGS 55A-16-20 obligates a nonprofit corporation to furnish a copy of its latest financial statement to any member who requests it. NCGS 55A-16-24 requires any nonprofit (VFD) which receives $5,000.00 or more in public funds to furnish its latest financial statement or a copy of its tax return (Form 990) to any member of the public upon request. However, the statute also allows the VFD to post a copy of the financial statement or tax return on a website in order to meet this requirement.

## Chapter 9

# Finance and Districts

## Introduction

This chapter will discuss the financing of municipal and volunteer fire departments and the creation and management of fire districts—two topics which are closely related. Since municipal departments in some parts of the state are relying more and more on volunteers and mutual aid from volunteer departments, a basic understanding of these topics is important. As discussed earlier, municipalities and counties may provide fire protection for their constituents (NCGS 160A-291 [and statutes following] and NCGS 153A-233, respectively). A municipality ordinarily provides coverage within the geographical boundaries of the municipality, the ETJ of the municipality, and areas covered by a contract with another entity. A VFD ordinarily covers areas described in contracts it has made with other entities. County fire departments ordinarily cover unincorporated areas of the county and areas which the county has contracted to cover. The designation of these areas of coverage frequently has a significant impact upon the finances of the fire departments involved.

A basic principle of contract law is that a contract does not exist (and become binding on the parties) until all parties agree to its terms. Municipalities, counties and VFDs would do well to keep this principle in mind when contracts for services are presented to them for approval.

# Acquisition of the Funding

## Municipal and County Departments

Municipal and county departments usually fund operations as items in their regular budgets, utilizing tax dollars collected from their constituents. However, because some municipal departments also enter into contracts with counties to cover certain unincorporated areas, they often are recipients of fire tax dollars collected from the covered areas (rural fire protection districts). NCGS 69-25.4(a) requires that the funds be used solely for fire protection. A frequently debated issue regarding these funds is the ratio of funds collected from the rural fire district versus the amount of fire protection funding budgeted and furnished by the municipality. The counties believe that the ratio should be based upon the value of the taxed property in the rural district versus the value of the property protected within the municipality and its ETJ. The municipal argument is that the ratio should be based upon the number of responses undertaken in the rural district versus the number of responses undertaken in the municipal coverage area. These are good arguments from both sides of the controversy, and the ultimate outcome of this as yet unresolved debate could have significant financial impact upon the parties involved.

Management of the municipal and county departmental funds is left to trained professionals working under the direct supervision of municipal and county managers and under the overall supervision of municipal and county boards. The rules of governmental finance are complex and lengthy and beyond the scope of this volume. Materials relating to municipal finance are available through the League of Municipalities, the Local Government Commission, and the Institute of Government (School of Government) in Chapel Hill.

A problem which occasionally arises in the municipal context is the acquisition by a volunteer component of a municipal department of funds of its own, typically as a result of fund-raising activities or donations. Many volunteer units have allowed these funds to be placed in accounts bearing the taxpayer identification number or name of the municipality. Once deposited, those funds now belong to the municipality and are subject to very strict (and harsh) rules regarding their management and expenditure. They no longer are under the control of the group which acquired them. Because of this problem, it is *vital* that volunteers in municipal departments establish mechanisms to manage any funds they acquire which they may wish to use as funds separate from those of the municipality. The author's home department had to deal with this problem in the fall of 1992. The emergence of the problem resulted in a lively confrontation between the municipality, the volunteers, and the members of the community served by the department before a solution was devised. The solution involved two actions. The first was the passage of a local act (House Bill 867, May 1993) which established rules for managing and disposing of the funds which had accrued to that date. The second part of the solution was the creation of a nonprofit corporation to solicit and manage funds donated on behalf of the volunteer component of the department. This nonprofit corporation, according to Section 501(c)(3) of the Internal Revenue Code, is qualified to manage the funds separately from the municipality—namely, as a charitable donee, allowing a donor to the corporation to deduct the donation from income taxes. This Section 501(c)(3) qualification, ordinarily accomplished by an amendment to the Articles of Incorporation, is vital for all VFDs as well.

Some FDs which also provide ambulance services as a part of their EMS services may have any additional funding source as provided in NCGS 44-51.1. Counties and municipalities can charge the patient for the services, and if the charges remain unpaid, may file a lien against the patient and attempt to collect the lien by court action (44-51.4). This is limited to municipal and county departments or agencies franchised by a municipality or county to provide the services and can be used only by departments located in the counties listed in NCGS 55-51.8. There are specific time-frames involved in this type of activity which must be followed precisely and the author recommends initiating such a program only after considering the possible political consequences of such an activity. Many taxpayers in North Carolina believe that their tax dollars are intended to provide such services and might not take kindly to being charged for such services.

## Volunteer Fire Departments

Traditionally, VFDs have relied on three sources of regular funding: donations, fire taxes, and direct payments from local governments. The steadily increasing duty requirements for fire departments and the skyrocketing costs of training and equipment have made donations only a small part of the source for operating funds for most VFDs. Most are now more dependent for funding on tax dollars, direct payment, and occasional grants. The basis for tax funding and direct funding is a contract between the VFD and a municipality or county. The contract sets out the basic relationship between the parties, stating, essentially, that in exchange for a certain amount of money, the VFD will provide fire protection services covering a certain geographical area. The contract should be examined carefully by the department before it is adopted. A contract is not binding upon the parties until all of them have agreed to its terms, so if the VFD discovers items in a proposed contract about which it has questions or objections, those items should be dealt with prior to approving and signing the contract. Consultations with legal counsel may be appropriate at this time.

Counties are increasingly requiring periodic audits of the finances of fire departments (including municipal departments) with which they enter contracts for services in an effort to account more precisely for the funds paid for fire protection. There is no statute which authorizes this requirement. The only statute which deals directly with nonprofit corporations receiving public funds (that is, tax dollars) is NCGS 159-40, which exempts several types of emergency service providers, including VFDs, from the requirement of an audit (NCGS 159-40(d)). As private nonprofit corporations, VFDs should be held accountable only to their members and constituents. Given these circumstances, the issue of audits should be negotiated between the parties to the contract. Audits are expensive, so any department faced with an audit issue in a contract for services would do well to negotiate the terms of, as well as the funding of, the audit prior to entering the contract. These negotiations may lead to an understanding between the parties which eliminates the audit requirement but which leaves in place a less costly mechanism of providing accountability. Municipalities are freestanding entities as well (municipal corporations) and are thus subject to rigorous financial oversight. The basic concepts for managing VFD funds are those set out in Chapter 55A of the General Statutes regarding good judgment and honest dealing.

If the VFD is accepting a direct payment from the county or municipality, as opposed to taxes, it is clearly in the department's best interests to establish a schedule for the payment of the money which best fits the financial needs of the department. This schedule should be established in the contract for delivery of services. Typically, this procedure will require some negotiation, because the entity supplying the money is using tax dollars to do so and tax dollars do not flow into a taxing authority on a continuous basis. They can be paid to the VFD only after they have been collected. Tax collections occur only during certain periods of the year, and the county or municipality has obligations other than fire protection which must be met, so payments for fire protection can become erratic unless a schedule can be set in advance by the parties to the contract.

A rural fire protection district (hereafter "fire district") is created by a referendum vote of the qualified voters living with the boundaries of the proposed fire district. Thirty-five percent of the resident freeholders (property owners) must first petition the county commissioners for a referendum election. Once the petition has been approved, a vote will be taken in accordance with the rules in Chapter 163A of the General Statutes to establish the district. If the vote is favorable, the district will be deemed established and taxation will commence. If the vote is unfavorable, another election cannot be held for two years (see NCGS 69-25.1). The maximum fire tax rate is $.15 per $100.00 valuation. The money collected can be used only for fire protection; however, the statute (NCGS 69-25.4) includes rescue services, EMS, and ambulance services as part of the definition of fire protection. The district can be abolished in similar fashion by vote of the resident freeholders (NCGS 69-25.10).

Fire district boundaries may be altered by several means (NCGS 69-25.11). One or more owners of property adjoining the fire district may request coverage (incorporation of their property into the district), and upon approval of the department involved, the county commissioners, and the district fire commission (if one exists), the area can be added to the district. Areas can be removed from a district in a similar fashion. When adjoining districts have similar tax rates, the departments involved, as well as any fire commissioners and the county commissioners, can arrange mutually acceptable terms for the boundary modification, publish notice of the proposed change in a local newspaper of general circulation, conduct a hearing, and then modify the boundary. When the districts involved have different fire tax rates, the proposed change must be submitted by a petition, reviewed by the various commissioners involved and by a public hearing, and then approved by vote of the appropriate commissions. If an area adjoining a fire district is located within the corporate limits of a municipality and is not afforded fire protection, the governing bodies of the entities involved may vote to include the area within the district. Occasionally the money collected from fire taxes is reallocated between adjoining VFDs by the county commissioners. This seems to occur when a portion of a fire district can be better served by an adjoining department. The taxes paid to the adjoining department will be those collected from the geographical area being covered. The county commissioners need only designate the department which is to provide the coverage. Altering the boundaries of a fire district simply is too complicated, in most instances.

When a municipality annexes part of a fire district, it removes the annexed area from the fire district, and the parties involved may enter such agreements regarding the disposition of equipment or delivery of services that they are able to negotiate between them (NCGS 69-25.14). The tax monies are to be prorated in accordance with the statute (NCGS 69-25.15). While a department which loses part of its tax base to annexation is entitled to compensation for the loss, there are specific timetables and other procedures which must be followed by the VFD involved in order to present a proper claim for compensation. These procedures are found in Article 4A of Chapter 160A of the General Statutes, beginning with NCGS 160A-37. In lieu of compensation, a VFD whose fire district is being annexed, either totally or partially, may negotiate with the annexing municipality to continue to provide services to the annexed area for five years. If this option is selected, the VFD must understand that strict time frames associated with these negotiations must be followed carefully. All documentation to be submitted should be examined carefully, possibly with the assistance of legal counsel, by the VFD whose territory is to be annexed. As long as local government budgets continue to be pushed to their limits by financial stress, it is vital that any VFD facing annexation follow these procedures precisely. If it does not do so, the VFD may be denied the compensation it seeks. It should be obvious to anyone that if a government can save budget funds by denying a claim against those funds, it will do so, thereby making the funds available for other projects.

A simpler and more flexible method to create and manage fire districts is for counties to establish service districts pursuant to NCGS 153A-301 (and the statutes following). Counties may establish districts to deliver fire services, rescue services, ambulance services, EMS, or other services. Typically, these districts are established when a service is not otherwise available. They are established by vote of the county commissioners and without the need for a referendum. Establishing service districts is a tool sometimes utilized to reallocate tax dollars collected within a fire tax district. Taxation is similar to that in fire districts, but sometimes with a cap of $.05 per $100.00 valuation for EMS services. This is frequently a method utilized by counties to add EMS services and taxation for them to an existing fire district. As with most new taxation, the commissioners are required to publish a notice and to conduct a public hearing before establishing the district. Territories added to a district are entitled to its services, but the residents must pay taxes (NCGS 153A-302). The reverse applies to territories removed from districts. The residents of the removed areas no longer are required to pay taxes, but they are also no longer entitled to services (NCGS 153A-304.1-307).

Donated funds should be accounted for in much the same fashion as other funds of the VFD. An important aspect of the solicitation of donations is the income tax status of the VFD. All nonprofit corporations are basically exempt from most forms of taxation, including income taxes. However, certain minimal requirements have been established by taxing authorities to monitor the activities of nonprofit corporations. One typical requirement is the filing of an informational income tax return, which discloses, in general terms, what has been done with the donated money. The filing of the informational tax return is essential to the continued nonprofit status of the corporation. The Internal Revenue Code (IRC) recognizes several categories of nonprofit corporation. They are found in Section 501(c). The most advantageous category for a VFD is the category described in Section 501(c)(3). To become classified as such, the VFD must state so in its articles of incorporation. If this was not stated at the time of incorporation, the articles should be amended to make this change in classification. By becoming classified as a "501(c)(3) corporation," the VFD becomes a more attractive potential recipient of donations, which can then be deductible from the income taxes of the donor. Occasionally these donations can be substantial—for example, land, buildings, equipment, and funds—all of which types of donations have occurred in North Carolina.

Increasingly, VFDs are seeking low-interest financing for projects—especially because of the skyrocketing costs of equipment. To the good fortune of VFDs, buildings are increasingly being financed at lower-than-market rates. Also, specialized financing companies and some commercial banks are entering this low-interest market. The United States Department of Agriculture, through the Rural Development Agency of the Farmers Home Administration, for example, has been in the low-interest loan business for a very long time. Other agencies, as well as the private business sector, also provide similar services. The granting of a low-interest loan to a VFD is dependent upon the fulfillment of two basic conditions: First, the VFD must establish the department's nonprofit status; second, it must establish its status as an agency acting "on behalf of" another entity. The nonprofit status is established by the articles and tax documents of the department, and the department becomes an "on-behalf-of" agency by providing fire protection on behalf of a governmental entity, either as a department of or as a contracting organization with a unit of local government. The objective of fulfilling the two basic conditions is to make the money loaned to the VFD nontaxable and the interest paid back to the lender largely nontaxable. By reducing the lender's tax consequences on the interest, the lender is able to reduce the interest charges to the VFD. Sections 265 and 150(c) of the IRC allow VFDs to be treated as political subdivisions of government for tax purposes, and, therefore, they benefit from low-interest rates afforded governmental bodies.

Section 150(c) of the IRC describes the requirements for a VFD to reach this status. The department must be a "bona fide" fire department organized and operated to provide fire and/or EMS services in a designated area which does not otherwise receive those services, and it must be required by a political subdivision to provide the services. Once these threshold requirements are met, the debt created by the VFD falls under Section 265 of the IRC as tax-exempt. As a duly designated tax-exempt political subdivision of North Carolina (for IRS purposes only), the VFD is now able to take advantage of the tax breaks available to governmental agencies.

However, utilizing these advantages carries requirements associated with purchases by governments. First, the project must be submitted for bids. This requirement can be troublesome because, usually, the project's specifications have been offered to a seller preferred by the department prior to the announcement of the project to the public. Ordinarily, this is a manageable issue if the department consults with someone experienced in such matters. Second, the VFD must conduct a public hearing to allow public input regarding its decision to proceed with the project. There are time frames associated with the hearing which must be followed carefully. Third, the project must be approved (or at least not objected to) by the government with which the VFD has a contract—a status which must be documented in writing. There are additional requirements which must be met, most of which either by the lender or by the governmental body (usually the county). The result is an extensive paper chase in which documentation of all requirements must be provided to the lender and/or county in order for the project to proceed. Lenders ordinarily require that a lawyer submit a written opinion regarding many of these conditions before approving the loan. The reader should be aware that the resulting legal services can be extensive, time-consuming, and, therefore, expensive. Clearly, it is in the best interests of the VFD to have as much of the necessary documentation as possible readily available for copying and delivery to those who need it. Time is money to a lawyer, so any time saved for the lawyer is money saved by the department.

## Spending the Funds

Departments owned and operated directly by governments utilize budgets carefully crafted, itemized, and adopted as ordinances. Expenditures are public record and subject to close scrutiny by various oversight agencies. Once the expenditure has been approved and the money acquired, the expenditure operates in accordance with standardized rules for the type of project involved and under the scrutiny of trained managers or finance officers and the governing board of the county or municipality.

The situation is more fluid when a VFD is making the expenditure. As a private nonprofit corporation, the expenditures are controlled by the decisions of the board of directors, corporate officers, and upon occasion, the members, and are not subject to public scrutiny. As pointed out earlier, Chapter 55A requires that the decisions be lawful, be made in good faith, utilizing sound business reasoning, and be made in accordance with the articles and bylaws of the corporation. Lenders sometimes will impose certain requirements regarding the expenditures, and these are often presented in the form of documentation to be submitted by the seller of the item(s) being purchased to the lender during various stages of the project in order to obtain payment. There can be other requirements, but these will vary, depending upon the nature of the project. As stated earlier, these expenditures are not subject to audit unless agreed to by the department. Often, expenditures associated with large projects will contain a consent to audit or an obligation to furnish a consent somewhere in the documentation. Departments would be serving their best interests considerably by carefully examining all documents associated with large projects before agreeing to their terms.

## Chapter 10

# Employment Law

## Introduction

Law relating to employment is complex, even when the focus is limited to the fire service and an in-depth discussion of the topic is far beyond the scope of this volume. Accordingly, this chapter will present an overview of the topic, leaving "signposts" available to the student for more in-depth consideration of specific topics as may be desired. North Carolina follows a basic rule of employment which sits at the heart of many decisions a department must make in the course of managing its personnel.

The rule is that North Carolina *is not* a "right-to-work" state. Unless the employer and employee contract for employment for a specified period of time, either the employee or the employer may terminate the employment relationship at any time for either good cause or for no reason at all. Employment is continued only at the will of the parties involved. This concept, known as "at-will," ordinarily will be modified by a written personnel policy created by the employer. Written personnel policies have resulted from a need to protect certain basic rights of employees required by either court or legislative action and to establish basic rules describing the ongoing relationship between the employer and the employees so that both sides understand their rights and obligations within the relationship.

## Hiring

Paid departments and VFDs ordinarily use different practices when hiring personnel. As a private nonprofit corporation, a VFD generally operates its hiring process in much the same fashion as any employer in the private sector, even though the services it provides are of a public nature. Ordinarily, the VFD is allowed to adopt hiring policies and procedures of its own creation. This allows it to recruit personnel in a manner best suited to its needs and financial capabilities. As will be seen later, however, there are certain basic rules relating to the workplace which all employers, whether paid or volunteer departments, must follow.

Recruitment is the first step in the hiring process. For a VFD, the process is a simple one: the department can "put the word out," by some means or another, that a job is available and await responses. The VFD can hire from within, without any external announcement, or it can follow any other announcement procedure it chooses.

A paid department is part of local government and must follow a different set of rules. Promotion within the department as a means to fill a position is pretty much left to the discretion of the chief or a designee, typically a committee. However, when recruiting from outside the department, basic rules must be followed. The employer must demonstrate objectivity in the recruitment process. Typically, this takes the form of advertisement in various publications available to both the public and the fire service community—online, newspapers and periodicals of general circulation and publications within the department which utilize some form of announcement available to all employees. Additionally, the position may be advertised with the Job Service of the Employment Security Commission. Job Service advertisement is required for all positions in state government but does not appear to be required for positions in local government. In accordance with Section 704(b) of Title VII of the Civil Rights Act of 1964, the advertiser cannot indicate any preferences based upon race, color, religion, sex, or national origin but can list any other job-specific requirements for the position (certifications, training, experience, and so forth). These requirements led in 1978 to the publication by the United States Government of "Uniform Guidelines on Employee Selection Procedures," which provides some basic guidance for employers in the recruitment process, hopefully avoiding some of the pitfalls which lurk within the rules for recruitment of personnel in public agencies. Once the potential employees have been located by recruitment, the next stage in the process is selection of the employees from within the group of applicants.

# Selection

Once again, VFDs have their own procedures for selection. As private non-profit corporations, they may develop their own criteria for selecting candidates from those available. Sometimes these criteria are developed by the chief, sometimes by committee, sometimes by the board of directors, sometimes by the membership, and sometimes by combinations of the foregoing. In any event, there should be some form of objective screening of the applications before selections are made. While an applicant may not be entitled to an explanation of why he or she was not selected, it is better for the public relations of the department to be able to demonstrate to the applicant who was not selected that the application was reviewed fairly prior to the selection of the candidate for employment.

In the public sector, the process of selecting an employee usually begins with an interview of each candidate, followed by a meeting with a group of interviewers. This group is increasingly becoming known as an "assessment board." The board is usually composed of qualified persons from within and without the department. All candidates for an open position are examined on the same set of criteria, and then one candidate is selected. The advantages of an assessment board are its objectivity and the variety of inputs afforded by having multiple members. It is an excellent means of avoiding some of the accusations of discrimination occasionally made by candidates who were not selected. The major disadvantages of an assessment board are the time and cost associated with assembling the board and transferring personnel from their ordinary duties to sit on the board. However, it is cheaper to pay for a board than to pay the costs of defending employment discrimination accusation-ask any trial lawyer. The board should be careful to ask similar or identical questions of all applicants. Federal law requires that public employers be objective and fair in their decision-making processes, and any procedure which contributes to objectivity and fairness helps preserve this attitude. All questioning should be governed by three basic principles recognized by the courts and other authorities:

1. Ask the candidates for information which will later be used in the selection process.
2. Ask the candidates job-related questions only.
3. Ask the candidates lawful questions only.

Questioning involving criminal convictions is one of the most sensitive topics in interviews. However, in the fire service such a topic is of vital importance. In order to protect life and property, the selected candidate will be required to enter private areas of peoples' lives while performing duties, sometimes without the consent of the owner or possessor, and will be operating large vehicles upon streets and highways of North Carolina, often at higher-than-normal speeds. These activities make a firefighter, in effect, a trustee of the public's well-being. North Carolina law imposes higher standards of conduct on people acting as trustees than on members of the public at large. It follows, therefore, that careful examination of the candidate for employment is necessary, if not mandatory.

Questions regarding marital status should not be asked unless it can be shown clearly and unequivocally that marital status has relevance to the job. Federal antidiscrimination law includes marital status as one of its sensitive areas. Questions about childbearing and changes in marital status can result in serious consequences to the department which uses those criteria in its selection of candidates.

Age is another factor to be regarded carefully. In most areas of public employment, age cannot be considered as an employment criterion unless the law specifically permits it. Both the fire service and law enforcement have age requirements regarding both the minimum age for employment and the age for mandatory retirement. Beyond these parameters, the interviewer should proceed very carefully.

Handicaps or disabilities should not be discussed. Both federal and North Carolina law forbid denial of employment solely on the basis of a handicap. However, like law enforcement, the fire service places physical demands on its personnel which are not found in other sectors of public employment, and therefore handicaps must be considered in the course of evaluating the candidate. An assessment board or interview is not the time or place to make the evaluation. There are other means of evaluating possible disabilities of the candidate. If the position being sought (for example, that of a dispatcher) imposes no special physical demands on the prospective employee, the department should evaluate all candidates objectively and make a selection. If the department can make reasonable accommodations for disabled personnel which are not unduly burdensome upon the department and which do not endanger other personnel, then it must consider the disabled candidates for the position, utilizing the same criteria as those used for candidates who are not disabled.

Political affiliation, religious preference, and organizational affiliation, including union activities, all are means of expression protected by both the First Amendment and the Constitution of North Carolina and are ordinarily not matters for consideration in the evaluation of a candidate. However, given the unusual nature of the services delivered by the fire service, organizational membership may well become a legitimate area for examination, especially since September 11, 2001. As a trustee of the public's well-being, the fire service has a duty to take whatever steps it can reasonably take to assure that its employees are worthy of that trust. After all, as of a few years ago, the person with the highest security clearance at Marine Corps Air Station, Cherry Point, North Carolina, an installation with some of the most advanced avionics in the world, was the fire chief. Think about it.

Once the candidates have been screened, a selection decision must be made. The courts have recognized the need for judgmental decisions in hiring. Ordinarily, it is not necessary for a decision to be made solely on the basis of pure data assembled in the course of the initial selection process. The instincts and impressions of the interviewers are recognized as legitimate tools to be employed in the selection decision. The secret to making the final decision is to utilize good judgment in a fashion likely to stand up well in court. Unfortunately, however, what constitutes good judgment is a matter to be considered on a case-by-case basis.

An extraordinarily thorny issue in the selection process is affirmative action, which is the use of one's race and gender as partial criteria for selecting suitable candidates for a position. Affirmative action was included as part of the Civil Rights Act of 1964 as a means to correct past injustices among women and African-Americans in employment and other matters, and it has been under almost constant legal attack since its inception as violating the equal protection clauses of the Constitution. Generally, the courts have upheld the doctrine. However, some states (for example, California) have voted to abandon it in certain contexts. In 2003, the United States Supreme Court began to impose limitations on the doctrine in two cases involving education. Until the doctrine is either abandoned or ruled unconstitutional, all public employers should be prepared to deal with issues created by hiring quotas. Hiring quotas may be created by court rulings or by the employer voluntarily. Court-imposed plans ordinarily offer written descriptions of what must be done. Voluntary plans, however, are created "from scratch" and can cause more trouble than they solve if not properly managed. While the courts have offered some guidance for creating a hiring quota plan, any such plan should be reviewed carefully with legal counsel before implementation. The courts have recognized four criteria to guide employers who are contemplating the use of affirmative action:

1. There should be a showing of prior discrimination.
2. The plan must be a reasonable attempt to remedy that prior discrimination.
3. The plan cannot adversely affect the rights of majority-category employees or serve as a complete bar to advancement for majority-category employees.

4. The plan must be a temporary one, intended to correct an existing imbalance.

Once the initial selection process has been completed, the department should make the offer *subject to* any additional physical, mental, or background investigative requirements. Screening by physical examination, physical capability testing, mental evaluation by appropriately qualified professionals, and background investigation should be explained carefully to the candidate at the time of the offer of employment. The department should also explain *that the offer of employment is conditional upon successful completion of this screening process.* It is vital that this notice be given to the candidate in writing and that the candidate acknowledge receipt of the notice in writing. A thorough investigation of the driving history of the candidate would seem to be appropriate as a beginning point, and, in any event, an essential part of the final selection process for all types of fire departments.

Homeland Defense issues have made careful review of potential fire service employees a vital concern for all departments. In 2003, the General Assembly passed an act to facilitate background investigations of fire service personnel by requiring that the SBI and the FBI furnish background information on potential members to any department which requests it. While well intentioned, the bill failed to address a number of issues relating to the cost of and the effects of such requests upon the managers of the databases from which the information would be collected. As a result, the implementation of the law has been put "on hold" until these issues are resolved. The statute is NCGS 114-19.12 and now allows any fire chief (paid or volunteer) to request the assistance. However, the funding and full implementation problems remain.

## Drug Testing and Searches at Work

Both drug testing and workplace searches are searches within the meaning of the United States and North Carolina constitutions and are controlled by the Fourth and Fourteenth Amendments of The Constitution and by Article I, Section 20 of North Carolina's constitution. The courts balance the nature of the intrusion into a person's privacy rights against the importance of the governmental interests being protected by the intrusion. In other words, is there an important governmental need which is sufficiently urgent or vital to justify a warrantless search of a person or property? The courts have recognized and accepted the idea of a warrantless search, but only in certain limited circumstances.

The two United States Supreme Court cases which first examined the concept of warrantless drug testing involved railroad workers and United States Customs Service personnel. The Court held that the United States Government had a compelling interest to protect the safe operation of interstate railways which was sufficiently vital to require warrantless drug tests. In the case of the Customs Service personnel, the Court pointed out that these personnel were a first line of defense against, among other things, illegal drugs, and that the government had a compelling interest in identifying personnel who might be using illegal drugs which exceeded the need for a warrant to conduct a test. Transportation and law enforcement personnel are clearly subject to random testing, whereas the random testing of other public employees is a requirement whose interpretation continues to evolve. Given the unusual nature of fire service duties and the increasingly wider response of the fire service to situations where national security is a major consideration, it can be argued that required periodic testing of all personnel is clearly allowable under the doctrine described above. Random drug testing can be made a part of any employment agreement, as well. Subsequent to September 11, 2001, required random testing may become the norm rather than the exception.

In a case in 1987, the United States Supreme Court held that, as a general rule, workers have a reasonable expectation of privacy under the Fourth Amendment in their offices, lockers, files, and other areas. In another case, a federal court held that fire chiefs have a lower-than-average expectation of privacy in their offices because the offices contain important records and equipment of their employers. These two rulings, combined with a reasonable suspicion of misconduct, justified a warrantless search of a fire chief's office. As a general rule, however, the courts will examine such searches on a case-by-case basis and balance the employer's need to supervise, control, and ensure the efficient operation of the agency against the right of the employee to privacy. Once again, however, Homeland Security needs may be seen by the courts as compelling justification for warrantless searches of the work areas of emergency services personnel. This issue continues to evolve.

## Firing

Volunteer fire departments ordinarily discipline their members in accordance with rules found in their articles or bylaws. However, the rules must meet certain constitutional requirements of due process. This means that the members must be informed of the accusations against them and afforded an opportunity to present their versions of the events that led to the accusations. In 1999 the North Carolina Court of Appeals, in *Wilson Realty v. Asheboro-Randolph Board of Realtors*, listed three procedures to be followed in order to eject a member from a voluntary association (for example, a VFD). First, the proceedings must be conducted in accordance with the organization's rules and laws. Second, the rules and laws must be consistent with public policy. Third, the member to be ejected must be given fair notice of a hearing relating to the discharge and afforded an opportunity to be heard in an atmosphere of good faith before an impartial hearing body (due process of law). Discipline of a paid member of a VFD is subject to similar due-process requirements. Because the paid member is an employee of a private nonprofit corporation and not subject to the State Personnel Act and because North Carolina is an employment "at-will" state, the VFD should proceed carefully when discharging a paid member, with the reasons for discharge thoroughly documented.

Ordinarily a paid firefighter, as a public employee, has a property interest in his or her job (wages, benefits, and so forth). This means that constitutional due process must be afforded the person to be disciplined. The employee must be informed of the accusations against him or her, be afforded an opportunity to prepare a response to the accusations, and be allowed to present a response to the accusations. The hearing on the accusations and the response must be conducted before an impartial decision maker. The employee should be afforded the opportunity to utilize the services of legal counsel, as well. Typically, the employer will have consulted counsel of its own if the discharge becomes contested. Additionally, the time, date, and location of the hearing must be given to the employee far enough in advance of the hearing to allow the employee adequate time to prepare any response.

According to the courts, probationary employees have no property interests in their jobs, acquire no due process rights, and, therefore, can be dismissed without notice and hearing. However, if the employee accuses the employer of a violation of Title VII of the Civil Rights Act of 1964 or the employee's First Amendment rights attach to the job, these rights must be protected as in the case of a paid public employee.

Public employees are entitled to appeal disciplinary rulings if they disagree with the outcome. Ordinarily, an appeal is made before the State Personnel Commission, whose ruling is only advisory to the local government employer unless the commission finds evidence of discrimination by the employer, in which case the ruling becomes binding. Either side in the dispute may appeal the ruling of the State Personnel Commission to the superior court and pursue the issue as far through the court system as possible.

# Family and Medical Leave Act of 1993 (FMLA)

The FMLA requires employers to provide up to twelve weeks of unpaid leave annually to employees who meet the eligibility requirements and it also forbids retaliation by employers against employees who request to utilize the rights afforded by the act. To be eligible, the employee must have worked for the employer for at least twelve months, must have been employed for at least 1,250 hours during the preceding twelve months, and must be employed at a work site where fifty or more employees are employed within seventy-five surface miles of the work site when the leave is requested.

The leave is available for the birth of and care for a newborn child; the placement of a child for adoption or foster care; care for the employee's spouse, child, or parent with a serious health problem; and care for a serious health problem of the employee. The leave can be taken intermittently or according to another schedule as long as a good-faith effort is made to avoid disruption of the employer's operations. The employer is required to maintain any health insurance benefits to which the employee would be entitled if the employee were continuing in employment during the FMLA period. The benefits may be discontinued if the employee announces an intention not to return from FMLA leave or fails to return after the end of the leave period.

Upon return from FMLA leave, the employee is entitled to the same position held prior to the leave, or to an equivalent position with equivalent pay, benefits, and other terms and conditions of employment. However, certain highly paid key personnel may be denied reinstatement if it is determined in good faith that the continued absence of the employee is likely to cause substantial economic injury to the employer's operations.

The FMLA requires that both the employer and employee carefully follow various procedures in order to utilize the benefits of the act. Both should examine these closely before utilizing it.

# Wages and Hours

## Basics

The general authority to set wages and compensation for municipal and county employees is found in NCGS 160A-162 and 153A-92, respectively. Both statutes allow the governing boards of the respective employers to fix the schedules of pay, allowances, and other compensation of their respective employees. In recent years, several cases have passed through North Carolina's courts relating to this authority, and the courts have held that the authority to establish the schedules for wages and other compensation lies exclusively with the governing boards. Typically, these cases have arisen out of circumstances when municipal or county managers have made unilateral modifications to pay and/or benefits without the approval of the governing body. The courts have held uniformly that these arrangements, however well intentioned, are void because of the specific authority set out in the General Statutes. In many instances, these issues involve overtime pay or supplemental benefits awarded to salaried employees as the result of services rendered in the course of responding to emergencies. Any attempts to modify compensation schedules by public-sector management without first obtaining the approval of the governing body concerned should be undertaken very carefully.

Two sets of laws control wages and hours in North Carolina—the Fair Labor Standards Act (FLSA), which is federal law, and the 1979 Wage and Hour Act of North Carolina, which is state law. Public employees (for example, firefighters) are covered by the FLSA. Employees of VFDs, while providing a public service, are, nonetheless, employees of private nonprofit corporations and thus appear to be covered by the North Carolina statute. The emergence of paid employees of VFDs is such a new issue that the law which covers them will be a matter for debate until amendments to the laws or court rulings resolve the issues created by this mode of employment. Currently, the trend by most VFDs is to follow the state guidelines, which require a shorter workweek for firefighters than the federal guidelines. The detailed regulations which govern the implementation of the state and federal wage-and-hour laws are complex and beyond the scope of this volume. The reader, however, can obtain copies of materials with more comprehensive discussions of these issues from the Institute of Government (School of Government) in Chapel Hill, the North Carolina Department of Labor (1-800-LABORNC), and federal labor authorities.

Both federal and state labor laws acknowledge two classes of employees—those subject to wage-and-hour requirements (nonexempt) and those who are not subject to the requirements (exempt), sometimes referred to as "blue collar" and "white collar" workers, respectively. The difficulty with the latter distinction (blue/white) is that the lines between those classifications have become blurred over time. Both laws have attempted to establish basic guidelines to identify exempt employees.

## Federal

The FLSA recognizes salaried executive, administrative, and professional employees as exempt. A "salary" is defined by 29 CFR 541.118(a) as a predetermined amount of pay per pay period which is not subject to reduction because of variations in the quality or quantity of work performed. Also examined with the salary issue is the employment situation of each of the three types of exempt employees listed above. Two tests, or sets of working conditions, are used to determine whether or not the person is exempt or nonexempt. These are referred to as "long" and "short" tests. Two tests are used because of the seemingly infinite variety of job descriptions and working conditions in the workplace and the need to provide a uniform set of criteria for determining an employee's status. The employee need only meet the conditions of one test in order to be classified as exempt. Because these tests are subject to periodic revisions, the conditions are not listed here. They can be found in the Code of Federal Regulations at 29 CFR 541.1 and 541.10 (executive), 29 CFR 541.2 and 541.20 (administrative), and 29 CFR 541.3 and 541.30 (professional).

The FLSA recognizes a distinction between firefighters and other employees when computing the hours in the employee's workweek. An ordinary employee must be paid overtime for any hours worked in excess of forty (40) during a typical workweek. A firefighter, however, must work in excess of fifty-three (53) hours during the same period in order to become eligible for overtime (one-and-one-half times the employee's standard hourly wage).

For many years, a bitter battle was fought within the labor law community regarding the proper classification of a firefighter for wage and hour purposes. Because many fire departments intermingle the fire and EMS duties internally, employers, on one side, were arguing that everyone in the fire department was a firefighter and therefore subject to a fifty-three hour workweek. The EMS community, however, argued that if a person's duties were primarily EMS rather than fire suppression or other duties traditionally acknowledged to be the responsibility of firefighters, the EMS personnel were entitled to the standard forty-hour workweek, regardless of their designations as firefighters. The U.S. Department of Labor attempted to remedy the problem by basing the classification upon the percentage of time in the workday an employee spent performing a particular classification of duty (the "80/20 rule"). This did not work well, either. In December 1999, President Clinton signed the *Fire and Emergency Services Definition Act* (HR 1693), which defines a firefighter for FLSA purposes. A firefighter is employed in fire protection activities and is trained as a firefighter, EMT, paramedic, rescue worker, ambulance crew member, or HAZMAT worker. Specifically, a firefighter (1) is trained in fire suppression, has the legal authority and responsibility to engage in fire-suppression activities, and is employed by a municipality, county, fire district, or state; and (2) is engaged in the prevention and extinguishment of fires or responds to emergency situations when life, property, or the environment are at risk. The bill makes no mention of percentages of time spent among the various duty descriptions.

While the definition of a firefighter for FLSA purposes is useful, it may not be the end of the story as long as employers continue to intermingle the duties of fire and EMS within their departments. The "80/20 rule" may not be as dead as Congress sought to make it.

Another aspect of the FLSA in dispute and only recently resolved is the question, What is a volunteer for wage-and-hour purposes? Section 203(e)(4) defines a volunteer as one who provides services for a public or charitable agency with no expectation of compensation. However, that person is not to be considered a volunteer if he or she is employed by the agency to which the services are provided. Translation: You cannot volunteer for your employer. Once the overtime threshold has been reached, any additional time worked, regardless of the intent of the work, must be paid as overtime.

Montgomery County, Maryland, is a leader in organizing and providing emergency services. They utilize a system of paid and volunteer providers organized under a countywide command-and-control agency. In the early 1990s, the International Association of Firefighters (IAFF) obtained a ruling from the U.S. Department of Labor which held that the organizational structure in Montgomery County was such that all paid firefighters worked for a single employer (Montgomery County) and, therefore, could not volunteer at any VFDs within Montgomery County without being paid overtime for their services. This ruling remained in effect until recently, when a series of federal court cases held that notwithstanding the presence of a central command-and-control agency, payment of tax funds to VFDs, mutual-aid agreements, county contracts, and other items, the volunteer emergency service providers were nonetheless independent agencies for FLSA purposes. This resulted in the rescission of the prior ruling and the resumption of a more typical relationship between the paid firefighters and the VFDs in Montgomery County.

It is apparent from this case that counties and municipalities must be careful with the terms of their emergency service contracts. A contract which intrudes too deeply into the operations of a VFD can establish an employer/employee relationship between the parties without intending to do so and thereby eliminate many of the cost-saving advantages created by the availability of paid personnel to volunteer at other departments.

## State

North Carolina's rules regarding the fire service are rather straightforward for paid firefighters: They are not covered by state regulations. Instead, they fall under the FLSA. NCGS 95-25.14 lists many other exemptions, including volunteers in nonprofit organizations (VFDs) where there is no employer/employee relationship (NCGS 95-25.14[a][1]). Otherwise, it appears that paid members of VFDs are covered by the North Carolina wage-and-hour rules. This coverage places them in the mainstream of employees, working standard forty-hour weeks, with no distinction between firefighters and other emergency services workers.

# Occupational Safety and Health (OSHA)

## Basics

North Carolina has had an OSHA program for governmental agencies since 1973, when the first version of NCGS 95-148 was enacted. However, the program did not acquire enforcement "teeth" until the tragic events in Hamlet, North Carolina, on September 3, 1991. The Imperial Foods fire demonstrated the desperate need for viable occupational safety and health and fire inspection programs on a statewide basis. Both programs are now in place, the former as a result of additional direct legislative action, the latter by the addition of a fire-prevention-and-safety code to the North Carolina building code in April 1992.

NCGS 95-148 is the statute which established the program. The North Carolina Department of Labor (DOL) is responsible for overseeing and enforcing the provisions of the program, and the commissioner of labor, through a designated deputy, directs the program. The state, together with all branches of local government, is required to acquire, maintain, and *require the use of* safety equipment and personal protective devices for its employees; to train the employees in the use of these devices; to submit annual reports regarding safety issues; and to keep records of occupational injuries and illnesses.

The statute *specifically exempts a VFD* from the OSHA requirements, provided that the VFD is not a part of a municipality. A municipality with a population of 10,000 or less may elect to exclude its fire department from OSHA by means of a resolution duly adopted by its governing body, but it cannot exclude any firefighters who are also employees of the municipality.

Fire Service OSHA standards differ from those of ordinary industry and are set out in a separate set of rules, written and published through the OSFM. These should be available through the DOL or the OSFM. Most of North Carolina's OSHA standards have been adapted verbatim from the federal rules. However, some of the rules in the fire service portion have been rewritten in North Carolina in order to more efficiently accommodate the needs of North Carolina's firefighters and fire departments.

An interesting aspect of this situation has been the question, What is a volunteer firefighter for North Carolina OSHA purposes? One part of the answer is membership. The firefighter must be a member of a VFD. Another part concerns money. Is the person in question receiving compensation (wages, salary, or the equivalent) for services rendered to the department? After a lengthy inquiry and debate spanning approximately four years, the Office of the Attorney General issued an opinion that such things as workers' compensation, insurance, death benefits, payments in reimbursement for expenses, reduced utility bills, and similar items were not, in and of themselves, types of payment which should be classified as wages or salaries for the purpose of OSHA.

The Attorney General also addressed the relationship between a VFD and the agency with which it contracts to provide services. Under certain circumstances, a VFD (and its members) could be deemed an employee of the local government agency with which it contracts to provide services. The focus of the opinion is control. Who directs the actions of the department, provides funding and controls its money, sets standards of performance, certification, or training, and owns its equipment and buildings? The Attorney General opined that control determines the existence or nonexistence of an employer/employee relationship between a VFD and the agency or agencies with which it contracts to provide services. If the agency exercises too much control over the activities of the department, an employer/employee relationship may be found to exist, resulting in the burden being placed upon the employer (municipality or county) to protect its employees (the VFD) at no expense to the employee (as required by OSHA). The Attorney General emphasized that the determination of any such relationship must be made on a case-by-case basis. Contracts for services should be scrutinized carefully in order to assure that controls not be imposed upon the VFDs activities which could create an employer/employee relationship.

The result of all the debate is that at this time, a volunteer firefighter is not under the OSHA umbrella as long as he or she is functioning as a volunteer. A person who is paid in Department A can volunteer at Department B, a VFD, and not be covered by OSHA. If Department B hires a few paid personnel during the day, for instance, the working environment of those paid VFD employees must meet OSHA requirements. Similarly, if paid and volunteer firefighters are working a scene simultaneously, the working environments of the paid personnel must meet OSHA requirements, while those of the volunteers need not do so.

## Workers' Compensation

Chapter 97 of the General Statutes (Workers' Compensation Act) establishes North Carolina's workers' compensation rules. As paid employees of local government agencies, paid firefighters are covered in much the same fashion as any other employees. Volunteer firefighters are entitled to much the same coverage as other workers when injured on duty. For a volunteer firefighter, the key to compensation is that the volunteer be engaged in activities ordinarily perceived to be duties, including responding to and returning from calls. Some activities are clearly duties or duty-related, but others, while undertaken as departmental activities, may not be so clearly defined. In a notable court case, a volunteer firefighter was held by an appeals court to be entitled to compensation for injuries sustained when he fell off a roof which he and other members of the department were repairing for a staunch supporter of the VFD. The ruling pointed out (quite correctly, in the opinion of the author) that the good will of the community was an essential part of a VFDs operations and that activities undertaken by the department in the furtherance of that good will fell within the compensation guidelines of workers' compensation. Generally, "on-duty" status is fairly clearly defined for paid personnel. It usually commences upon arrival at work and terminates at the end of the shift. Arguably, paid personnel responding to an emergency recall or "all-hands" recall are covered from the time they initiate a response to the recall, since they have been placed "on duty" as of the time of the recall.

Unfortunately, while workers' compensation does pay many of the costs associated with the injury, it does not pay all of them. Certainly, it is in the best interests of most departments and firefighters that they or their organizations maintain additional insurance to cover some of the "gaps" in workers' compensation coverage whenever it is financially feasible to do so.

The notice requirements of the workers' compensation statutes should be followed carefully in order to protect the viability of a claim. Like any other insurance company, a workers' compensation carrier is in business to make profits and generate dividends for its shareholders—not to pay claims. Accordingly, if the necessary documentation is late or missing, the company will very likely deny the claim.

# *Woodson v. Rowland* 329 N.C. 330 (1991)

In August 1991, the Supreme Court of North Carolina handed down its ruling in *Woodson v. Rowland.* Mr. Thomas Sprouse was killed in a trench collapse in Durham County, and his estate filed a wrongful death lawsuit against his employer and other companies involved in the construction project in spite of the existence of Workers' Compensation coverage. The evidence showed that Sprouse was ordered into an unprotected trench by his employer; that his employer had prior knowledge of the hazards involved and had ignored them previously; that the employer was present at the scene and observed the hazards; and that others' opinions and scientific evidence showed that the trench and soils were unsafe.

The court ruled than an employer engaged in an inherently dangerous activity who intentionally disregards safety factors, orders his employees into dangerous circumstances, and as a result injures or kills an employee may be held accountable in spite of the presence of Workers' Compensation coverage. The court found that such behavior amounted to intentional injury to the employee and, therefore, the employee was not restricted to Workers' Compensation coverage when seeking compensation for his or her injuries. The court held the individual supervisor (Rowland) personally liable, as well as his company. Rowland was a subcontractor on the job, and the court held that the primary contractor, Davidson & Jones, could not delegate responsibility for safety for employees engaged in inherently dangerous activities to a subcontractor. Davidson & Jones could thus be held liable as well. The incident commander (IC), in a mutual-aid context, should keep this ruling in mind when organizing a response. The presence of an active and forceful safety officer is a good idea under these circumstances. This concept of the liability of the primary contractor (or the department requesting the mutual aid) has been strengthened, not weakened, by later rulings by the courts.

In the original 1991 case, the court did rule, however, that one who pursues one of these options must eventually choose between the Workers' Compensation settlement and the results of the lawsuit, though both options can be pursued simultaneously. Obviously, the choice of remedy would have to be a careful one, based upon the circumstances of each case. Cases of this type are now often referred to as "Woodson" claims.

In a subsequent case, the court ruled that an employee who acted on his or her own and disregarded safety rules was not entitled to pursue a "Woodson" claim. Fortunately, as of this writing, a "Woodson" claim against a fire department has not reached an appellate court for a ruling.

In 1996, the City of Charlotte was sued for wrongful death because defective radio equipment had led to the death of a city police officer. The estate of the officer sought to pursue a "Woodson" claim because Charlotte appeared to be insured. However, the Supreme Court of North Carolina ruled that the risk-sharing arrangement to which Charlotte belonged did not amount to insurance, that sovereign immunity applied, and that therefore the "Woodson" claim could not continue. In 2001, the court ruled that merely alleging an OSHA violation on the part of an employee would not be sufficient to create a basis for a "Woodson" claim. There would have to be more misconduct on the part of the employer before such a claim could be allowed.

As a result of various "Woodson" cases, the Supreme Court of North Carolina has identified six factors which must be considered by a court in evaluating such a claim:

1. Did the risk involved in the harm exist for a long time without causing injury?
2. Was the risk created by a defective instrument with a high probability of causing the harm at issue?
3. Did the employer attempt to remedy the risk that caused the harm prior to the accident?
4. Did the employer's conduct which caused the risk violate state or federal safety regulations?
5. Did the employer create a risk by failing to follow the industry practice for the activity in question, even if there were no state or federal violations?
6. Did the employer offer training in safe behavior "appropriate in the context of the risk causing the harm"?

It is apparent that successful pursuit of a "Woodson" claim will be an uphill fight because of the complexity of the legal analysis necessary to make the case successful. However, given the inherently dangerous nature of most fire service activities, intentional disregard of safety factors in fire-ground or training evolutions could have consequences for a department and/or its command structure reaching far beyond mere Workers' Compensation claims. The doctrine created in the *Woodson* case is as applicable to the fire service as it is to general industry, and it is something which should be kept in mind by those who plan and conduct training and fire-ground evolutions.

## Sexual Harassment

Title VII of the Civil Rights Act of 1964 prohibits discrimination on the basis of, among other things, gender. Gender discrimination is a sensitive topic in the workplace today, and accusations should be investigated carefully and violators of the policy punished appropriately. Coping with sexual harassment issues is a matter of risk management, which translates itself into training and understanding. The burden has been placed upon the employer (and subordinate supervisors) to establish and enforce an effective anti-sexual-harassment policy.

Sexual harassment is defined as comments, actions, or things sexual in nature which a reasonable person finds offensive. Rules resulting from court decisions and various regulations can be used to avoid the events which lead to sexual harassment complaints. There should be a clearly stated policy prohibiting sexual harassment—with a "zero tolerance" clause. All members of the department must understand that they must treat each other with dignity and respect. There must be regular training on the policy and the training must be documented. Supervisory personnel must enforce the policy. Creation of the policy is best left to experts, and fortunately, like many personnel documents, sample policies are available to assist a department and its designated expert.

Until recently, sexual harassment claims were deemed to be nonenforceable if no damage resulted from the incident (the "no harm, no foul" rule). In order to be held accountable, the person causing the harassment would have to be seeking to achieve a particular goal rather than simply acting in a fashion perceived to be obnoxious. Requiring a sexual favor in exchange for a promotion or a privilege would be harassment, but obnoxious behavior not constituting criminal conduct would not be harassment.

However, two cases illustrate how the definition of "sexual harassment" has changed and how difficult it can be to deal with this issue. In 1998, the United States Supreme Court took away the "no harm, no foul" rule. In one case involving a lifeguard and in another involving a worker in general industry, the Court held that an employer may be held liable for the consequences of a sexually hostile work environment created by a supervisor of the victim. These cases suggest that an entire chain of command supervising a victim can be held accountable if appropriate action is not taken to deal with the situation and that simply creating a sexually hostile environment, without any quid pro quo activities, can be sexual harassment subject to compensation for a victim. In a case from the Supreme Court of North Carolina (*Parikh v. Eckerd Corp.*), the court held that boorish behavior did not create sexual harassment unless the behavior affected a term, condition, or benefit of the victim's employment. The facts of this case make interesting reading and the author wonders how carefully the North Carolina court considered the 1998 United States Supreme Court cases.

On 28 June 2019, a civil lawsuit based on allegations of sexual assault and sexual harassment entitled *Jane Doe v. Bedford County, Virginia and John W. Jones, Jr., Chief of Department, Bedford County Department of Fire & Rescue* (Case No. 6:19CV00043) was filed in the United States District Court for the Western District of Virginia. This case is an important one for us in North Carolina because of where it was filed and because of its subject-matter.

Filed in federal court in Virginia, this case falls under the federal Fourth Circuit Court of Appeals which means that should this case go to trial and subsequently be appealed, the opinion issued by the Fourth Circuit Court of Appeals could have an effect on sexual harassment laws and policies in Maryland, West Virginia, South Carolina, Virginia, and North Carolina. The facts of the case are instructive to North Carolina emergency services providers because they illustrate what can happen to a department and its supervisors when they fail to have or fail to enforce rules or policies dealing with sexual abuse and/or sexual harassment.

The circumstances which led to the filing of this case occurred in February, 2018 when a 52-year-old male training officer (Hawkins) employed by the Bedford County Department of Fire & Rescue (Department), forced his sexual attentions on a 17-year-old EMT trainee who was under his supervision (Jane Doe, the Plaintiff in the lawsuit). The Hawkins subsequently was charged in criminal court in Virginia and of this date (July, 2019), is serving a sentence there.

The Plaintiff had applied for membership as a volunteer EMT with the Huddleston Life Saving & First Aid Crew, Inc. (Huddleston), an emergency services provider operating in Bedford County, Virginia. In order to so serve, she had to become trained and certified as an EMT by the Commonwealth of Virginia. Part of her training consisted of a mandatory series "ride-alongs" conducted by Field Lieutenants employed by Department. She was assigned to Hawkins for that purpose. Huddleston was not involved in this aspect of her training.

Hawkins was employed by Department in 2007 and subsequently promoted to Field Lieutenant. A criminal background check was performed on Hawkins by Bedford County at the time of his application and it revealed that he was a convicted violent felon.

The lawsuit argues, among other things, that the Chief's failure to train his employees regarding sexual harassment, assignment of a convicted felon to "ride-along" duties with minors, and failure to take disciplinary action against Hawkins constitute deliberate indifference to the Plaintiff's constitutional right not to have her person attacked and violated by governmental employees and/or agents.

The legal basis of this case is violation of the Plaintiff's rights as protected by the Fourteenth Amendment as applied through the Civil Rights Act (42 USC 1983), protection from the abuse of a private citizen by government or its employees or agents. Plaintiff has demanded Thirty Million Dollars ($30,000,000.00) in damages from the defendants Jones and Bedford County.

This case is an object-lesson for all departments and supervisors of why having a sexual harassment policy, training personnel in the sexual harassment policy, and *enforcing its provisions* is so important.

# Chapter 11

# Benefits

## Introduction

This chapter discusses benefits available to firefighters and fire departments in North Carolina. Funds for benefits come from federal and state sources and, unless otherwise indicated, are available to both paid and volunteer departments or their members. The law controlling these funds is found in the United States Code or the North Carolina General Statutes, with any supporting rules in the Code of Federal Regulations (CFR) or the North Carolina Administrative Code (NCAC). The rules in the CFR and the NCAC should be examined whenever an application for benefit funds is considered.

In 2012, a financial study of relief funds requested by the General Assembly was completed. The study, in turn, triggered a reconsideration by the General Assembly of the then-current rules governing the management and disbursement of relief funds. After extensive discussions and negotiations by and between the General Assembly, OSFM, and the NCSFA, the General Assembly enacted Session Law 2014-64, amending Article 84 of Chapter 58 relating to local relief funds and Article 85 of Chapter 58, relating to statewide relief funds. Among the amendments were name changes for the Articles, making them much easier to identify. This chapter will discuss both the old and revised versions of the two Articles.

# North Carolina Benefits

## Firefighters Relief Fund

This benefit is funded by monies collected from fire insurance premiums paid in North Carolina every year. NCGS 58-84-1 requires all insurance companies to report all premiums paid for fire and lightning coverages to the state treasurer by March 15 of each year. After being taxed in accordance with NCGS 105-228.5(d)(4), the net proceeds are credited to an account at the DOI for disbursement.

NCGS 58-85-25 sets out the rules for disbursement. Three percent of these monies and an amount equal to what would be paid to any fire districts which are not members of the North Carolina State Firefighters' Association (NCSFA) are paid to the NCSFA. Two percent of these monies are retained by the DOI to cover administrative costs. The remainder of the funds are paid to fire districts which are members of the NCSFA in proportion to the level of activity in each district (as required by NCGS 58-84-50). When the money reaches the fire district, it must go into a relief fund administered by five trustees.

The recipients of funds from the DOI are identified by district rather than department because of the occasional overlap of district and fire department response areas.

Trustees are discussed in NCGS 58-84-30. The board of trustees for the relief fund is composed of five members—two members elected by the fire department with primary responsibility for serving the district, two members selected by the local governing board of the AHJ, and one member selected by the Commissioner of Insurance. All members serve staggered terms of two years each in order to prevent total replacement of the board at any single time. The fire chief or chiefs of the department or departments are ex officio members of the board if they are not selected to serve as trustees. The trustees must be bonded, and the costs of the bonding are usually deducted from the monies paid into the fund by the DOI.

Competent record keeping is an essential part of the relief fund administration process. NCGS 58-84-40 requires that records of receipts and disbursements be furnished by the trustees to the DOI by October 31 of each calendar year. Because eligibility for relief funds is conditioned upon membership in the NCSFA, the NCSFA must furnish certifications of its members to the DOI in January of each year (NCGS 58-84-40(b)). Also, the governing bodies of the fire departments must furnish rosters of their NCSFA-eligible members to the NCSFA by December 31 of each year (NCGS 58-84-46). Failure to do so usually results in the return of annual relief fund payment to the state treasury.

Should a department experience difficulty having its roster ready by December 31, it may obtain an extension of time by contacting the NCSFA. However, a department would be ill-advised to make extensions of time an annual habit.

If irregularities are discovered within the relief fund, the statute requires that all additional funding of the fund be withheld until the NCSFA has conducted an investigation of the irregularities satisfactory to the DOI and has ruled upon the issues presented. If criminal activity is suspected, law enforcement agencies will become involved. Investigators from the DOI, the Banking Commission, the SBI, the FBI, and any other state or federal agency with jurisdiction can participate.

Sometimes, a relief fund becomes depleted because of claims or low funding. The statute allows the fund to borrow money from the state as long as the funds are repaid at the statutory rate of interest (typically 8 or 9 percent). Borrowing money to replenish a depleted relief fund is not considered good management—except in an extreme emergency.

Once obtained, how can the funds be spent? NCGS 58-84-35 gives fairly specific guidance. They can be spent to protect an active-duty firefighter, or a dependent thereof, from loss sustained as a result of a line-of-duty (LOD) injury. They can be spent to assist a destitute firefighter who has at least five years of service if the assistance is approved by the NCSFA. Funds can be used to pay membership assessment for the Firemen's Fraternal Insurance Fund of the State of North Carolina. They can be used to supplement benefits paid under the Firefighters' Relief Fund (FRF) and provide educational assistance for firefighters or their dependents who otherwise qualify for the FRF. However, supplemental or educational payments cannot be made unless the NCSFA certifies that such payments will not endanger the relief fund making the payments. This certification should be obtained in writing and made a part of the permanent record of the relief fund involved.

Under North Carolina law, a trustee is classified as a fiduciary. This means that the trustee is perceived to be of good character, worthy of the trust of others, and capable of honestly managing the assets of others. A fiduciary is expected to carefully protect the assets in his or her custody and exercise his or her best judgment in managing them. A fiduciary can be held accountable for irresponsible management of assets under his or her control in civil court and can be held accountable in criminal court for the theft of those assets. Such a crime is called a crime of moral turpitude, and a conviction of a person having committed the crime can ruin that person's future employment prospects by preventing him or her from obtaining a job which involves significant responsibility. Relief fund trustees should bear this in mind when making their decisions.

Unfortunately, relief fund irregularities are not uncommon in North Carolina, and this should be a warning to those with oversight over such funds to follow the statutory guidelines provided and consult with the DOI or the NCSFA when questions arise regarding the use of the funds. Ordinarily, a trustee who abstains from a vote regarding the use of the funds will not be absolved of blame if a problem arises. North Carolina courts regard an abstention as an affirmative vote even though it is not counted in the total. If a trustee disagrees with a proposed decision, he or she must be sure that his or her vote is recorded as such in the minutes of the meeting. If a trustee suspects that the proposed action is criminal in nature, the trustee, as a fiduciary, must report that activity to law enforcement authorities. By failing to do so, a trustee is violating his or her duties as a fiduciary and may be held accountable as an *accomplice* to whatever crime is being committed.

## (2014) Article 84 Local Firefighters Relief Fund

In NCGS 58-84-5, the definition of a fire district for the purposes of Articles 84 through 88 of Chapter 58 is modified and a fire district is now considered to be a political subdivision of the State of North Carolina if it has an organized fire department, with equipment of a value of at least $1000.00, which has been rated and certified by OSFM, and which has a response area approved by an agency of local government (municipality or county). It should be noted that rating requires at least a "9" under the current rating system and that the department submit an annual roster to OSFM and NCSFA containing the requisite number of firefighters possessed of at least 36 hours of documented annual training. NCGS 58-84-25 remains essentially unchanged and the provisions of NCGS 58-84-30 relating to trustees for the local relief fund was not modified. However, given the recent extensive scrutiny given to the local relief fund system, departments should review their trustees and their selection process in order to be certain that the necessary standards and procedures are being followed.

NCGS 58-84-32 adds a more demanding standard for the management of relief funds. The trustees are now designated trustees within the meaning of NCGS Chapter 36E, which means that the relief fund trustees now have a statutory standard against which their decisions and behavior will be judged. The standards for decision-making are those of a prudent trustee as set out in NCGS 36E-3. The funds must be managed for the purposes for which the fund was created with the following conditions taken into consideration:

1. General economic conditions;
2. Inflation and deflation;

3. Tax consequences to the fund, if any;
4. The effect of a decision upon other funds held by the fund;
5. Expected return on the investment;
6. Other financial resources in the fund;
7. The need to preserve capital.

Decisions must be undertaken in consideration of what effect the decision will have on all of the assets of the fund. However, there is no restriction on the type of investment the trustees might select, but any such decision should be made while keeping the seven items listed above in mind. Diversification of investments is required of prudent trustees, but in the case if relief funds, it may be impractical because relief funds are intended to make cash available on very short notice in order to assist someone in need. Receipts and disbursements are made as cash transactions by check and liquidating an investment in order to pay someone's rent or utility bill, for instance, only prolongs a process which is intended to be relatively speedy.

General standards of care for the management of trust funds are good faith and the behavior of an ordinarily prudent person in a similar position under similar circumstances, taking into consideration the seven factors listed above. Relief fund trustees have been granted some immunity under NCGS 58-84-40 for some of their decisions as long as one of the five offenses listed in the statute has not been committed. Because the statutes are so new, presently we do not have any court cases to offer any guidance on proper behavior by relief fund trustees, so the best path to follow would seem to be one which follows the written rules as closely as possible.

Disbursement of funds at the local level has undergone major modifications. In NCGS 58-84-35, the General Assembly changed "fireman" to "firefighter" throughout the statute. Subsection (2a) of Subparagraph (a) requires that the financial problems to be addressed by the fund not be the fault of the Applicant for relief and permits assistance with housing, vehicle/commuting expenses, food, clothing, utilities, medical care, and funeral expenses. Subsection 5 allows payment for Pension Fund and workers compensation premiums in addition to previously approved insurance or pension premiums if the firefighter is otherwise qualified for benefits under the Firefighters Relief Fund. Subsection 7 permits payments for annual physicals and Subsections 1, 2, 3, 4, and 6 remain essentially unchanged. There are limitations on these payments which should be examined and followed carefully by any fund considering making them.

Subsection (b) of NCGS 58-84-55 imposes limitations on how and when certain expenditures can be made under Subsection(a) of the statute. If the fund is deemed "financially unsound" many otherwise allowable expenditures must be approved in writing by the NCSFA before they can be made.

Detailed accountings of annual expenditures and a full accounting of qualifications of all department members who are eligible for benefits must be certified by the trustees of the relief fund to the NCSFA by October 31 of each year (NCGS 58-84-40(a)) and the NCSFA must report this information to DOI by January 1 of each year. If mismanagement of funds is identified, the relief fund will not receive any additional funds until the problem is corrected. The data will be managed using FD identification numbers issued by OSFM to each FD (NCGS 58-84-41). If no report is submitted by a department by January 31, the department's relief fund will not receive any funds for that year and the funds not paid to the department's fund will be paid into the statewide relief fund (NCGS 58-84-46). Another addition to Article 84 is NCGS 58-84-52 which allows individual firefighters whose departments are not members of NCSFA to be paid North Carolina line of duty death benefits, but only if the funds are available in the forfeited funds held in the statewide relief fund.

There are some departments which have obtained local acts to allow them to do certain things with their relief funds. NCGS 58-84-65 repeals most of those local acts in an effort to get all department operating their relief funds under a uniform set of rules. Any department currently operating its relief fund under a local act should examine this statute closely with the assistance of its attorney to assure that it is not in violation of the new relief fund rules.

Another important addition to the relief fund rules is NCGS 58-84-33. It begins by setting a cap on the amount of funds which can be held in a local relief fund—an amount equal to $2500.00 for each member of the department on its roster (NCGS 58-84-33(a)) on January 1 of each year. Subsection (b) obligates the NCSFA to inform DOI of the permissible amount for each department. If the department's fund exceeds the permissible limit, the department will not receive funds for the year in question. However, Subsection (d) of the statute allows a department to dedicate a portion of its relief funds to a supplemental retirement fund for its members. If this program is established, it must be managed under the supervision of NCSFA, and if the funds are used solely for supplemental retirement purposes, the funds so dedicated to the supplemental retirement program will not be counted as part of the allowable maximum fund balance specified under Subsection (a) of the statute.

## Appropriated Relief Fund (Statewide Firefighter's Relief Fund)

This is a statewide relief fund created by NCGS 58-85-1 and administered by the NCSFA. Funds may be appropriated by the General Assembly and paid to the treasurer of the NCSFA. They are controlled, administered, and disbursed by the NCSFA. To be eligible to receive funds, the firefighter/applicant must be a member of the NCSFA in good standing.

The triggering event for eligibility is a line of duty (LOD) death of or injury to an eligible firefighter. The benefit is payable to a firefighter, a surviving spouse of the firefighter, the child or children of the firefighter, and/or a dependent mother of the firefighter; and it can be utilized only after local relief funds are exhausted. The fund can also be used for academic scholarships for eligible recipients, for supplemental premium payments to the Fraternal Insurance Fund on behalf of firefighters above the age of sixty-five years, and to provide accidental death and dismemberment insurance coverage for firefighters whose departments are not eligible for relief-fund monies. The statute prohibits lawsuits to enforce its provisions, placing the decisions regarding eligibility and payment in the sole discretion of the NCSFA.

For the purposes of this fund, "line of duty" means fire-service duties and any other duties directed to be performed by a fire-department officer in charge. The statute does not define "officer in charge," but it seems reasonable to assume that the term is intended to include any firefighter in charge of an activity of the department.

NCGS 58-85-20 defines members of the NCSFA as members of organized fire departments and individual firefighters who otherwise meet the membership requirements of the NCSFA.

In 2014, the General Assembly made some changes to the statutes relating to the firefighters' statewide relief fund in Article 85 of Chapter 58. They began by renaming the chapter to make it easier to identify by calling it "Statewide Firefighters Relief Fund." NCGS 58-85-1, as amended, leaves the administration of the fund with the NCSFA, but the General Assembly made two important modifications. First, they deleted the requirement that a local relief fund be exhausted before a firefighter becomes eligible for state benefit funds. The second change allows, but does not require, the NCSFA to award academic scholarships and provide certain insurance benefits to any firefighter, regardless of his or her membership status with the NCSFA. It seems obvious that any the availability of any of these benefits would be contingent upon the availability of funds and therefore not guaranteed. In addition, the NCSFA is required to follow prudent funds management practices as required by NCGS Chapter

36E, with limitations on certain expenditures which are similar to those imposed on local relief funds.

## State Fire Protection Grant Fund

This fund was created within North Carolina's Office of State Management and Budget pursuant to NCGS 58-85A-1(b). It is intended to compensate local government for the costs associated with the protection of state-owned buildings and other state-owned property located within the boundaries of a local government AHJ. The statute authorizes the General Assembly to appropriate funds in the following amounts from the following sources:

a. $3,080,000.00 from the General Fund,
b. $150,000.00 from the Highway Fund,
c. $970,000.00 from receipts from the Consolidated University of North Carolina, and to pay them to eligible applicants.

The allocation and distribution of the funds is limited to AHJs which have state property supported by the General Fund, the Highway Fund, or the Consolidated University of North Carolina Fund which in need of fire protection located within their respective jurisdictions. The Office of State Management and Budget should be consulted regarding details of eligibility and is distribution.

## North Carolina Firemen's and Rescue Squad Workers' Pension Fund

This fund was created pursuant to NCGS 58-85-1 in 1957 to provide pension payments and other benefits in order to encourage participation in fire service and rescue activities. Paid and volunteer participants are eligible for membership.

The fund is managed by a board of trustees consisting of the state treasurer, the Commissioner of Insurance, and four members appointed by the governor (NCGS 58-85-5). The trustees utilize the general management powers and duties specified in NCGS 58-85-10. They are required to appoint a director of the fund, who is responsible for the day-to-day administrative operation of the fund (NCGS 58-86-15). The custodian of the fund is the state treasurer, who is charged with the responsibility of investing the monies in the fund and keeping those monies in a specifically identified fund. (Some years ago, a bill was introduced in the General Assembly by members who professed to be friends of the fire and rescue services, to remove the fund from its current location to the state's General Fund, where the monies could be used for any purpose whatsoever. It took concerted and diligent lobbying efforts by various groups—especially the NCSFA—to prevent the transfer of the funds.)

Basic criteria for eligibility for benefits under the fund are set out in NCGS 58-86-2(5). The person must be a firefighter or a fire marshal of the State of North Carolina or one of its political subdivisions and must belong to a bona fide fire department. The department must provide protection rated at a minimum of 9, A, or AA by a North Carolina-recognized insurance-rating system, must own at least $5,000.00 worth of apparatus and equipment, and must hold at least four hours of meetings and drills (training) each month. The firefighter must complete at least thirty-six hours of meetings and training each calendar year. NCGS 58-86-2(5) was amended, effective January 1, 2015, to eliminate the "meetings" portion of the statute. Eligibility for membership in the NCSFA now requires 36 hours of training each year. This requirement can be prorated for any new member of a department based upon the date when the new member joins the department. These activities should be documented *very carefully* by the department, and the member should retain a copy of any such activity, especially training, should the training records be lost or damaged. The statute limits a department whose members would be eligible for the fund to twenty-five persons, plus one member per 100 population served. For example, a VFD serving a district with a population of 10,000 people could have a maximum membership of 125 volunteers (25 plus 100). The obvious reason for the cap on membership is to prevent departments from loading their rosters with personnel simply seeking to benefit from the fund. The governing body of each department with members participating in the fund must furnish to the NCSFA by December 31 of each calendar year a roster of persons eligible for membership in the fund. This list must, in turn, be delivered to the state treasurer by the following first day of July. The reporting of this roster is also vital because it establishes the eligibility of the persons listed thereon for other important benefits.

NCGS 58-86-2(13) defines "Training" as an activity which, if attended by the department member, "will result in the preparation of, or knowledge gained by, the member in the areas of fire prevention, fire suppression, or protection of life and property." The rule further requires that the training be conducted for the "purpose of providing a learning or preparation experience for the members."

Eligible rescue-squad workers are persons who are members of bona fide rescue- or EMS-service providers, who are otherwise eligible for membership in the North Carolina Association of Rescue and EMS, Inc., and who can document at least thirty-six hours of training and meetings in the preceding calendar year. As with firefighters, a roster from each eligible department must be reported in order to assure eligibility for this benefit as well as others.

Presently, the monthly contribution required of each member of the pension plan is $10.00 (NCGS 58-86-35,40). The monthly benefit payable to those eligible to receive it is $170.00 (as of 2019)—a rather remarkable return on the investment, considering that some of the recipients of pension-plan payments had paid only $5.00 per month during their terms of service.

To receive payments, the applicant must have completed twenty years of service during each of which he or she has documented at least thirty-six hours of meetings and drills (training) and has paid into the fund an amount of money equal to the sum of 240 monthly installments—another illustration of the importance of an accurate annual roster submitted to the NCSFA.

Retroactive membership is also available to otherwise eligible participants in the fund (NCGS 58-85-45). The fund is opened periodically for repurchase of prior unpaid eligibility at the then-existing monthly rate (presently $10.00) plus interest or at an actuarially computed figure. A person under thirty-five years of age who is not a previous member of the plan or who joined and did not receive credit for the prior time may join the plan and purchase prior time at the current monthly rate plus interest. Other applicants may be able to purchase prior time, but the payment will be computed actuarially and can be expensive. However, considering the return on the investment, the payment may well be worth the added expense. NCGS 58-86-45 and the Pension Plan Office of the State Treasurer should be consulted before attempting to purchase prior time because sometimes the issues presented can be complex. There also is an administrative charge for joining the fund, as provided in NCGS 58-86-50.

The monthly benefit amount is set by the General Assembly, based upon budgetary data furnished by, among others, the state treasurer. This figure and the monthly membership payment are set out in NCGS 58-86-55, which also discusses retirement for purposes of the fund. The member must have completed twenty years of service, paid the necessary monies into the fund, and reached the age of fifty-five years to receive payments. This is applicable to both career and volunteer members of the fund.

A member who is disabled in the line of duty may begin to receive payments at age fifty-five without making any further payments into the fund subsequent to the determination of disability. A member who is disabled outside the line of duty and who has at least ten years of eligible service may continue to make payments until the requirement of 240 payments has been met; then he or she may begin receiving payments at age fifty-five.

Occasionally, departments with eligible members are put out of business because of annexation or absorption by another entity. An otherwise eligible member of the pension plan who no longer has a means to serve may continue to make monthly payments up to the 240 threshold and then begin receiving pension plan payments at age fifty-five.

Under certain circumstances, lump-sum payments are made by the fund (NCGS 58-86-60). When a person is no longer eligible for membership or withdraws from membership, the monies paid into the fund will be returned to those persons or entities that made the payments. The monies paid into the fund are paid to the family of a member who dies before age fifty-five, and if the member dies after beginning to receive benefits, the fund will return the amount paid into the fund, less any amounts paid out in benefits.

Should the fund become depleted, NCGS 58-86-65 allows the state treasurer to reduce the monthly benefit payment amount pro rata in order to adjust for the depletion. NCGS 58-86-70 empowers the General Assembly to alter the fund and pension plan at any time, with or without the consent of the participants. The statute states that no member of the pension plan has any vested property right in the plan. This provision appears intended to prevent any participant from filing a successful lawsuit regarding any disagreement over some aspect of the pension plan or the fund.

The trustees of the fund are authorized to determine what constitutes creditable services for the purpose of pension-plan eligibility. They are authorized to award credit when a worker may be denied credit for some reason not of his or her making (NCGS 58-86-75). A worker need not serve his or her entire twenty years with a single provider. The key to eligibility is to be sure that each unit with which the worker served has documented the service (NCGS 58-86-80). Delinquent payments can lead to forfeiture of membership in the fund. A member who misses six consecutive months of payments can have his or her eligibility terminated (NCGS 58-86-85). This requirement places the burden on the worker to pay attention to his or her fund records in order to prevent an unfortunate surprise at a later date.

Pension plan funds are exempt from most types of financial liability—for example, from judgments and bankruptcy but not from Equitable Distribution and the provisions of NCGS 110-136. This exception should be watched closely by any members of the fund contemplating a separation or divorce which may involve a property settlement (NCGS 58-86-90). Fund members who belong to state or public school employee associations with 20,000 members or more may authorize deductions from the fund payments to be paid to the associations (NCGS 58-86-91). However, these payments are terminated automatically if the association to which the payments are being made attempts to engage in collective bargaining activities.

## Volunteer Fire Department Grants (NCGS 58-87-1)

Created by the General Assembly, placed within the DOI, and managed by the state treasurer, the Volunteer Fire Department Fund is intended to provide matching grants for VFDs. The monies in the fund are to remain in it, rather than being moved around among other items in the state's budget, thereby assuring the continuing availability of the monies. The Commissioner of Insurance is responsible for distributing the funds and, ordinarily, manages the applications. The statute requires that the commissioner, insofar as is possible, distribute the grants evenly around the state. Each applicant must match the amount presented in the grant request with an equal amount of its own funds, and each grant is limited to a maximum of $30,000 per year. The use of the money is limited to capital expenditures—for example, building construction and equipment purchase. Compensation for personnel, for instance, would not be an acceptable reason for a grant application. VFDs which are parts of municipalities are also eligible for the grants. The key to eligibility appears to be the basic organization of the department. Whether the VFD is managed by a board of directors or a governing board of a political subdivision of the state appears to be immaterial. The basis for selection of a recipient is need.

To be eligible, the VFD must be staffed entirely by volunteers; it cannot have more than the equivalent of six full-time paid members; and it must be certified by the DOI as a fire department.

All grants are to be awarded on May 15 of each grant year, with the results reported within sixty days to the General Assembly. These time frames make it vital for an applicant to plan carefully in advance in order to avoid losing a grant because of tardiness. Most grants are very specific regarding time frames and application documentation, making it difficult for an applicant who has made a mistake with the time or documentation to mount a successful challenge to a denial of the grant.

NCGS 58-87-1(a1) was amended in 2014, to be effective January 1, 2015. An important change was implemented to assist small-budget FD's by changing the scale for matching funds. If the FD receives less than $50,000.00 annually from an agency of local government, the matching amount is now $1.00 from the FD for every $3.00 of grant funds. For example, if the FD receives $40,000.00 in funding from a county and it applies for $15,000.00 in grant funding, the FD would be required to pay $5,000.00 in matching funds because the amount received from local government is less than $50,000.00 that year.

Additionally, grant funds can now be used to put into service property acquired from the Department of Defense. For example, converting a military truck into a tanker or brush truck.

## Volunteer Rescue/EMS Grants (NCGS 58-87-5)

Also established by the General Assembly, the statute creates a fund to provide matching grants to volunteer rescue units or volunteer rescue/EMS units. Use of the grant monies is limited to equipment purchase and capital improvements. To be eligible, the primary duty of the applying unit must be rescue or rescue/EMS. Application is to be made to the DOI, which establishes the eligibility criteria. The Department of Health and Human Services is to furnish a list of equipment eligible for funding. The monies are held by the state treasurer for investment in accordance with the law, with the earnings remaining in the fund—again, another attempt by the General Assembly to protect the use of the funds. The grants are to be made on December 15th of each year.

A recipient, who must use nonstate funds to match the grant, can receive up to $25,000 per grant; and a recipient who is not required to use nonstate funds to match the grant is limited to $3,000 per grant. If the identified liquid assets of the recipient exceed $1,000, the recipient must match the grant dollar-for-dollar. This requirement makes the vast majority of the grants dollar-for-dollar in the same fashion as the fire department grants. Recipients are limited to one grant per fiscal year.

To be eligible, the applying department must be staffed entirely by volunteers, except for paid staff constituting not more than the equivalent of ten full-time employees; must be recognized by the DOI as a bona fide rescue or rescue/EMS service provider; and must be declared eligible for a grant by the Department of Health and Human Services (approved by OEMS). It also must meet the requirements of subsection (a) of the statute relating to dollar-amount eligibility.

NCGS 58-87-5(c) contains, among other things, useful definitions of *rescue* and *rescue squad*. The statute requires that the term *rescue* be used to designate an emergency-services provider only if it has been certified as a provider of rescue services by the North Carolina Association of Rescue and Emergency Medical Services, Inc. *Rescue squads* or *units* are defined as organizations whose members are not necessarily firefighters, EMS workers, or law-enforcement personnel but those who are trained to remove persons from dangerous situations to locations of relative safety. Modifications similar to those in NCGS 58-87-1 were made to NCGS 58-87-5, as well.

NCGS 57-87-7 has been modified to require on-site review by DOI of the use of grant funds for a period of up to five (5) years after every grant is received. This is a very important operational consideration for every grant recipient because it places the burden upon the FD to account for all equipment or training purchased with the grant funds. Subsection (b) of the amended statute requires the FD to reimburse the State of North Carolina for the value of any equipment acquired by grant funds and disposed of within the five year period, less depreciation. This makes it very important for the recipient of the grant to account for or insure (if possible) any equipment so acquired in order to avoid having to reimburse the State for any lost equipment. Subsection (c) of the statute requires any department which ceases to exist to transfer all grant-acquired property to the FD which takes over coverage for the now non-existent department or "... to another appropriate department that may effectively use the equipment."

## Volunteer Safety Workers' Workmen's Compensation Fund (NCGS 58-87-10)

The statute creates an expendable trust fund within the DOI which is intended to be freestanding and to retain all the proceeds of the investments made by the fund. VFDs and volunteer rescue/EMS units which are not part of local government and which are exempt from state income tax under NCGS 105-130.11 are eligible. The monies in the fund are used to provide workers' compensation benefits to members of eligible units in accordance with the existing rules for such payments as set out in Chapter 97 of the General Statutes (the workers' compensation statute).

The State Fire and Rescue Commission administers the fund through a third-party administrator hired by the commission and without the contracting requirements of Article 3C of Chapter 143 of the General Statutes. The commission has the authority to adopt its own rules to govern its relationship with the fund's administrator.

Revenue for the fund originates from two sources. The first is direct appropriations to the DOI by the General Assembly. The second source is payments in a fixed per capita dollar amount by a participating unit, based upon the roster it has submitted. The amount of the payments must be sufficient to enable the fund to pay benefits to its members as required, and payments must be made on or before July 1 of each year. Payments are to be made to the State Fire and Rescue Commission, which transfers the funds to the state treasurer, who acts as custodian of the funds.

This statute was amended by the 2014 legislation. Paid and volunteer members of eligible organizations are now eligible for these funds (NCGS 58-87-10(c)). The statute now requires annual actuarial studies to be performed by DOI to better manage funding of this program. The data generated by the actuarial study is furnished to State Budget and Management for calculation of the amounts necessary for annual funding. When the calculations are completed, the information is given to the Secretary of Revenue, who remits the required amount to the program (NCGS 58-87-10(g)).

## Rescue Squad Workers' Relief Fund (Chapter 58, Article 88)

This fund is controlled by the DOI, with the executive committee of the North Carolina Association of Rescue and Emergency Medical Services, Inc. serving as the fund's board of trustees. The fund can be used for the following purposes:

1. To protect rescue or EMS workers in active service from financial loss because of job-related sickness or injury;
2. to provide financial support for members who have died in the line of duty;
3. to provide academic scholarships for the children of active-duty, retired, or deceased members;
4. to pay death benefits to dependents of members who have died in the line of duty; and
5. to pay such additional benefits as may be allocated.

Membership is open to both individuals and organizations. Individuals who can document their participation in at least thirty-five hours of meetings and training each year are eligible for the benefits, and units which can certify to the DOI by January 1 of each year that all their members meet the individual eligibility requirements are also eligible for membership. Any eligible member of the fund who incurs an illness or becomes disabled in the line of duty is eligible for benefits. *Line of duty* is defined as rescue- or EMS-duty only, and not some other job-related activity (NCGS 58-88-10).

To seek a claim for benefits, a chief officer of the claimant's unit must submit written notice of the event(s) triggering the claim to the fund within thirty days after the event(s) and must request the necessary forms to justify the claim. The forms justifying the claim and a certification of the eligibility of the member shall be submitted under oath (or affirmation) by the certifying officer. These forms must be accompanied by a physician's certification regarding the medical condition of the claimant. Once payments of benefits have been initiated, the recipient must file an annual justification form with the fund as long as the need for the claim exists.

Filing a false or fraudulent claim under a benefit program is a serious criminal offense which can rise to the level of a felony. Typically, the felony is obtaining property by false pretense, which can carry a penalty involving substantial imprisonment. Since this type of felony is a crime of moral turpitude, conviction of such a crime usually ends a career because it implies that the person convicted is not trustworthy. The nature of fire, rescue, and EMS duties requires that those performing them be extremely trustworthy because they are often involved in some very private aspects of the lives of those whom they serve. Since the preservation of trust and confidence is essential to the continued effectiveness of emergency services, crimes of this nature involving emergency services workers have been dealt with harshly by the courts.

## North Carolina Death Benefit (NCGS 143-166.1, and Statutes Following)

This benefit is payable only upon the occurrence of a line of duty death (LODD). It is payable to the designated survivor(s) of a firefighter who, while performing official duties, is killed, dies as a direct result of injuries, or dies as a result of extreme physical activity or exercise which are in the scope of official duties. Eligible personnel are firefighters who meet the criteria set out in NCGS 58-86-25 and DOI employees who are engaged in firefighting activities, training other personnel, or serving as members of State Emergency Response Teams (SERTS). Payments are made to the spouse, dependent children, or dependent parents of the decedent, as the case may be.

Official duties are defined as activities en route to, engaged in, or returning from training and activities while responding to, engaging in, and returning from a call. A firefighter who acts on his or her own volition because it is reasonably apparent that prompt action is necessary to protect persons or party and that any delay in response would worsen the property damage or further endanger human life, is also eligible for the benefit.

Twenty thousand dollars is payable at the time of death, to be followed by annual payments of $10,000.00 thereafter until a total of $50,000.00 has been paid. The rules for administration and payment of the fund are made and implemented by the Industrial Commission. The statute forbids any appeals of rulings in these matters beyond those appeals available within the Industrial Commission (NCGS 143-166.4). This benefit has no effect upon any other payable benefits (NCGS 143-166.5) and is tax exempt (NCGS 143-166.6).

## North Carolina State Firefighters Association Death Benefit

As of 2005, the benefit pays $60,000, plus an additional $10,000 per dependent child. The department of the deceased member also receives a $6,000 bereavement benefit payment. A recipient of the benefit must be a member of the NCSFA at the time of the death.

## Miscellaneous

Traveling to and from training exercises may be covered by the LOD benefit plans as a "line-of-duty" activity if the training is properly documented by the department in question prior to attendance by the person in question. Arguably, this documentation should include written records of the proposed training, including rosters and specific authorization of or direction for attendance by a person in the department with the authority to issue such an authorization or order. Fire service instructors face an uncertain status unless they are working with their home departments within the scope of their ordinary duties. In order to help guarantee coverage, departments should include instructor activities as part of the ordinary duties or job descriptions of those members who will engage in instructional activities. When the instructor has contracted through the Department of Community Colleges (DCC) to conduct the training, the instructor is an independent contractor as far as the DCC is concerned, and it appears that the instructor might be covered by the benefit or Workmen's Compensation only if he or she is working with his or her home department. At the time of this writing, there do not appear to be any court cases which offer any guidance on these issues.

# Federal Benefits

Currently, two major benefits are available through federal sources. One is a death benefit and the other is intended to upgrade the equipment and training of those departments who are delivering the services. The death benefit is known as the Public Safety Officers' Benefit Program (PSOB) and the equipment and training funding program is known as the Assistance to Firefighter's Grant Program (AFG).

## Public Safety Officers' Benefit Program (PSOB)

This benefit is payable to firefighters who are killed or permanently disabled in the line of duty. It is administered by the United States Department of Justice (DOJ), and as of January 2019, it's estimated benefit could be in excess of $350,000. Until recently, a highly controversial aspect of the benefit was that it did not pay if a firefighter died or was disabled as a result of a stroke or heart attack, unless the stroke or heart attack could be attributed directly to specific activities undertaken in the line of duty.

On 15 December 2003, Congress, by House Bill #919 and Senate Bill #459, enacted the Hometown Heroes Survivor Benefits Act. The bill originated in the House of Representatives, cosponsored by, among others, Representatives Bob Etheridge and Howard Coble of North Carolina. The bill modifies the PSOB to add heart attack and stroke as acceptable (covered) causes of death if the following conditions are met:

1. The firefighter is on duty; and
2. the firefighter is engaged in a nonroutine stressful or strenuous fire suppression, rescue, HAZMAT, EMS ... or other emergency response activity; or
3. the firefighter participates in a training exercise involving nonroutine stressful or strenuous activity; and
4. the firefighter dies of a heart attack or stroke suffered while engaging in an activity or while on duty subsequent to said activity *or* not later than twenty-four hours after the conclusion of said activity; and
5. All the above presumptions cannot be disproved by competent medical evidence.

There can be little doubt that this is an improvement over the earlier law; however, one aspect of the rule should be kept in mind. Item 5 above leaves the door open for the administering authority (the DOJ) to contest any heart attack or stroke claim utilizing its own medical evidence or to delay payment until any perceived medical issues are resolved. Obviously, the law is too new for any interpretations of these rules to be generated. Therefore, they should be watched closely, pending further developments.

## Assistance to Firefighters Grant Program

This program is administered through the FIRE Act. Recently, the funds for the program were moved from the USFA to the Department of Homeland

Security (DHS) in its Office of Domestic Preparedness. The administrative aspects of the program remain managed by the USFA, with funding from the DHS.

The focus and intent of the funding is the enhancement of the basic firefighting capabilities of fire departments nationwide. Currently, documentation guidelines and timetables are being followed strictly because of the huge volume of applications for assistance. Strict adherence to such guidelines and schedules is a legally acceptable practice for screening purposes—a matter all applicants must consider when preparing their requests for assistance.

Several methods have been available to search for information regarding applications for assistance. The United States Fire Administration (USFA) can be contacted by telephone at 1-866-274-0960 and at three websites: www.fema.usfa.gov, www.usfa.fema.gov/grants/, and www.ojp.usdoj.gov/dop.

Applications for assistance to purchase apparatus are very difficult to obtain unless extreme hardship by the applying department can be demonstrated. Currently, properly documented applications for specialized training or smaller items of equipment (for example, turnout equipment or SCBA) are receiving favorable responses.

An important requirement of any FIRE Grant application is a certification by the applying department that all of its members have been trained in the then-currently federally approved system of incident management. Failing to meet this requirement likely will result in either denial of the grant application or attempts by DHS to reclaim any funds provided under the grant. Intentionally falsifying an application could result in criminal prosecution.

## Chapter 12

# Hazardous Materials (HAZMAT)

## Introduction

Regulations concerning hazardous materials may be the most complex body of law to be confronted by the fire service. It involves a convoluted interaction of both state and federal statutory law, together with regulations published in the CFR and NCAC. In the ordinary course of events, when there is a conflict between state and federal law, the federal law or rule will control the situation unless there exists a specific rule which hands the authority to the state. A detailed discussion of these statutes and regulations is well beyond the scope of this volume. The purpose of the exhaustive training undertaken by those who manage HAZMAT evolutions is to familiarize and continually update them with the latest versions of these statutes and regulations. This chapter will discuss the basics of North Carolina HAZMAT rules as they relate to the fire service, emphasizing the rules and regulations which every fire department should keep in mind, regardless of its level of HAZMAT response training or capability. All members of the fire service who participate in HAZMAT responses must bear in mind that no matter how annoying a particular rule may appear to be, it was enacted to protect both the responder and the public. Each rule or regulation imposes a duty upon the responder to comply with its specifications. Modifying or ignoring a rule can have unpleasant legal consequences for the parties involved if injury to persons or property results from the disregard or modification of the rule or regulation. However, if a better way of accomplishing the mission can be found, the rule or regulation should not be ignored. If the responder is able to construct an improved "mousetrap," the new device or procedure should be brought to the attention of appropriate authority and *advocated vigorously* so that an alternative, improved procedure can be made available for general use.

An excellent example of this idea is the approach taken by OSFM regarding live-burn exercises. North Carolina subscribes to the basic safety standards for live-burn exercises promulgated by NFPA 1403. However, as any instructor who conducts live-burn exercises will tell anyone who will listen, trying to get a meaningful (instructive) room full of fire by using pallets and straw is generally a waste of time. Accelerant is an essential aspect of a truly instructive live-burn exercise. OSFM recognized this problem and modified North Carolina's training standard to permit the use of safe and reasonable amounts of accelerant. The result, as the writer happily testifies, is far more instructive for the student.

Because HAZMAT evolutions and live-burn exercises both involve higher-than-normal risks, it is logical that any improvements in existing "mousetraps" should be presented for possible approval and widespread implementation. Since, historically, the fire service has been forced to rely on innovation to accomplish missions which it has been handed but not always with the proper equipment (and HAZMAT certainly is one of these missions), the fire service should avail itself of any opportunities to present new ideas.

# North Carolina HAZMAT Law

Most of North Carolina's HAZMAT statutes are found in two chapters of the General Statutes. Article 18 of Chapter 95 provides rules for identifying hazardous substances, and the North Carolina Emergency Management Act (Chapter 166A) provides mechanisms, and some rules, for conducting responses to HAZMAT events.

## Identification (Hazardous Chemicals Right to Know Act)

Much of the oversight and identification responsibility for HAZMAT, when it is located within North Carolina, belongs to the DOL through its Occupational Safety and Health Division (OSHNC). Hazardous substance lists (HSL) and material safety data sheets (MSDS) are created and administered through OSHNC. The Hazardous Chemicals Right to Know Act contains definitions for terms used in it, including *fire chief* (senior official for the local fire department) and *fire department* (the fire department with jurisdiction over the event or facility where the event occurs).

In NCGS 95-191, the act requires that any possessor of HAZMAT who stores or maintains certain quantifies thereof maintain a Hazardous Substance List (HSL), listing and describing what substances are located at the site in question. All containers of chemicals must have labels which are not defaced and all HAZMAT must be labeled as such. The basic rule for quantities is fifty-five gallons or 500 pounds, whichever is greater. The Act also requires that the chemical or common name of the substance be shown as well as the maximum amounts of Class A, B, C, and D chemicals stored, utilizing guidelines related to the quantities of each substance. The HSL must also show the area where the chemicals are stored and whether or not there are storage requirements related to temperature and/or pressure. The HSL must be updated at least quarterly, if necessary, and never less than annually, because of changing conditions related to the chemicals. If quantities change significantly, or chemicals are added or deleted from the list, the HSL must be amended no later than thirty days from the change. The custodian of the chemicals is also authorized to substitute a federal form from within the Superfund Act in lieu of the HSL. Regardless of the form used, it is clear from the wording of the statute that the custodian of the HAZMAT must maintain an accurate list of the HAZMAT in its possession.

An MSDS must be provided to each manufacturing and/or nonmanufacturing purchaser for each hazardous substance purchased from a manufacturer or distributor. An employer—for example, a fire department—with possession of any such chemicals must possess an MSDS for each chemical for which it has custody. If the MSDS does not arrive with the chemicals, the employer is required to request one in writing within thirty days of the date of arrival of the chemical. If the request is not answered satisfactorily within 30 days of the date when it was submitted to the manufacturer or distributor, the matter is to be referred to the DOL for appropriate action.

A matter of great importance to the fire service is emergency-response information. NCGS 95-194 sets forth some rules in this regard. Subsection (a) requires that the custodian of fifty-five gallons or 500 pounds of HAZMAT furnish, *in writing*, to the chief of the department with jurisdiction over the materials the name of and contact information for a representative of the custodian who can be contacted for additional information, together with a properly prepared HSL. The HSL is to be properly updated and all updates furnished to the chief. The custodian is also required to furnish a copy of an MSDS to the chief upon written request.

Any department with jurisdiction over HAZMAT must conduct preplanning in order to organize appropriate responses to any HAZMAT emergencies.

NCGS 95-194(c) directs the custodian of the HAZMAT to admit a fire chief or his or her representative into the premises of the custodian for the limited purposes of preplanning responses and verifying the accuracy of any HSL furnished by the custodian. The request for the visit must be submitted to the custodian in writing prior to the visit. The statute is silent regarding how far in advance the request must be submitted. However, a court would describe adequate notice as that which is reasonable under the existing circumstances.

An interesting question is raised when the fire department is refused entry after making the appropriate request. Certainly, the department could go to court and ask for a court order, known as an injunction, directing the custodian of the HAZMAT to allow entry. However, such activities are both time-consuming and expensive, but, until a better method of dealing with the problem is made available, possibly some form of administrative search warrant, a court order may be the only effective recourse.

The fire chief with jurisdiction is authorized to request, in writing, that the custodian of the HAZMAT develop a written emergency response plan for the facility which contains the HAZMAT. The plan must contain evacuation procedures, a list of the emergency response equipment available at the site of the HAZMAT, and any other emergency response plans associated with the HAZMAT. The custodian of the HAZMAT must furnish a copy of the emergency response plan to the fire chief with jurisdiction if asked to do so in writing (NCGS 95-194(e)).

If the fire chief receives the emergency response plan, the chief must make the plan and all applicable MSDSs available to the members of the FD with jurisdiction over the HAZMAT, together with other emergency response personnel with responsibility for planning and conducting an emergency response to a HAZMAT event. The emergency response information obtained by the chief must not be disclosed by its custodians to those who lack the "need to know," and its use must be limited to preplanning an emergency HAZMAT response. Intentionally disclosing the emergency response information is a Class 1 misdemeanor and is punishable as such. As an additional protection for the information, it is exempted from the Public Records Act's requirements that the information be made available to the public as a publicly maintained record (NCGS 95-194(f, g)).

If someone wishes to file a complaint regarding the violation of some aspect of the act, it must be filed in writing with the DOL; and any complaint filed by a fire chief must be investigated by the DOL without any unnecessary delay. Upon presentation of proper credentials, an investigator from the DOL is entitled to a right of immediate entry to the facility which is the subject to the investigation. When the investigation has been completed, its results are to be presented to the DOL for appropriate findings. If either side in the matter disagrees with the findings, that party may request an administrative hearing to review the findings in order to uphold, overturn, or modify them. If violations of safety regulations are found, penalties may be assessed by the DOL against the custodian of the HAZMAT. The penalized party may also appeal any adverse rulings through the administrative appeals process and then through the courts (NCGS 95-195).

In its investigative process the act also contains a whistle-blower statute. NCGS 96-196 prohibits an employer from firing, disciplining, or discriminating against an employee who brings a problematic situation to the attention of appropriate authority. This protection should be brought to the attention of any employee being interviewed during the course of an investigation.

Trade secrets must be handled carefully by emergency responders. Ordinarily, trade secrets are allowed to be retained by an employer and not disclosed to parties outside the group of employees with the need to know. However, the courts have long recognized that the need to protect the public is superior to the need of an employer to protect a trade secret. NCGS 95-197 is an attempt to achieve a middle ground between protection of the public and protection of trade secrets. A custodian of HAZMAT who believes and claims that the HAZMAT is in fact a trade secret may keep the information as long as it is furnished to the fire chief with jurisdiction. Once received by the chief, the information must be held in confidence.

The statute allows *any person in North Carolina* to request that the DOL conduct a confidential review of the claim of a HAZMAT custodian that the information is protected as a trade secret. If the DOL determines that the claim is not completely valid, the ruling can be appealed. If the claim is determined to be valid, the DOL must determine whether or not the information disclosed to the fire department up to that point is sufficient to permit the department to formulate an appropriate response plan. If the DOL determines that the information furnished up to that point is insufficient, the DOL must order the HAZMAT custodian to release sufficient information to the fire department for it to formulate a response plan. This ruling, likewise, is appealable.

Custody of the information requires care. Any custodian of the information, including the DOL, who knows the information is confidential and discloses it can be charged with a Class I felony and is also liable in civil court for monetary damages and/or an injunction on behalf of the HAZMAT custodian whose information has been disclosed improperly.

To summarize, the statute requires that HAZMAT information be disclosed even though it may, in fact, constitute a trade secret. However, once disclosed, the information must be protected carefully and its use restricted to those persons with a need to know. Inappropriate release of the information can lead to serious criminal charges and civil liability.

Another exception to the confidentiality rule relating to trade secrets concerns medical emergencies. A HAZMAT custodian must disclose immediately to treating medical personnel any necessary chemical information required for emergency medical treatment. Once disclosed, the custodian may impose any reasonable confidentiality requirements upon the medical personnel who receive the information. A written request for the information is not required (NCGS 95-198). In nonemergency situations, the information must be disclosed to responsible medical personnel when the request is submitted in writing stating the need for the information. The releasing party may request that the receiving party sign a confidentiality agreement prior to the release of the information. If the recipient of the request for information refuses the release, or fails to do so after thirty days from the request, the DOL must conduct an investigation to determine whether or not there exists an enforceable trade-secret claim.

Communities have been given a right to request HAZMAT information as well. NCGS 95-208 authorizes *any person* to make a request if certain conditions are met and procedures followed. If the request is refused, or not answered within thirty days, the DOL must conduct an investigation. The custodian of the HAZMAT does not have to disclose the information if it can be shown that the information is a trade secret or that the request is being made by or on behalf of a competitor or that the requestor failed to comply with the requirements of the statute authorizing such a request.

Chemicals transported in interstate commerce, used for personal consumption, held in retail food and trade establishments, used as food additives, food colors, and cosmetics, and held in laboratories under the supervision of properly qualified personnel are exempt from the act. A farming operation which employs ten or fewer full-time personnel is exempt as long as it routinely stores less than fifty-five gallons or 500 pounds of HAZMAT. Distilled spirits, tobacco, and untreated wood products also are exempt from coverage, as are medicines used for patients in health care facilities or for experiments in laboratory facilities.

The act forbids units of local government from requesting disclosure of information which would otherwise be protected under the act. The General Assembly wanted to create a uniform, statewide system for managing HAZMAT information. Two options remain open to local government, however. The one described in the statute is the adoption by a unit of local government of a nationally recognized fire code. Arguably, a municipality or county which modifies its building code by the adoption of nationally recognized fire-code provisions requiring that otherwise protected HAZMAT information be disclosed might execute an "end run" around some of the protective language in the act. The second option can occur when a HAZMAT custodian is seeking approval of an activity through a planning board or board of adjustment. The board could make disclosure of the HAZMAT information a part of the review and approval process. A fire chief who pays attention to what is taking place within the department's jurisdiction should discuss the matter with the reviewing board and impress upon them the importance of the information.

## Responding (North Carolina Hazardous Materials Emergency Response Act)

This act, consisting of NCGS 166A-20 through NCGS 166A-28, was created to establish a system for regional HAZMAT emergency response within North Carolina (NCGS 166A-20(b)). Oversight of the program is given to the Department of Public Safety (DPS), with management of the program supervised by its Division of Emergency Management (DEM). The act requires that the program be coordinated, wherever possible, with other emergency responders across the state. The program was established to include at least six strategically located HAZMAT-response teams with the capability to be dispatched on a 24-hour basis by DEM's Emergency Operations Center. The DPS is required to work in consultation with the Regional Response Team Advisory Committee when establishing the program. The Committee is appointed by the secretary of the DPS and consists of one representative from each of the following: the DEM, the NCSHP, the State Fire and Rescue Commission, the DEHNR, the DOT, the Department of Agriculture and Consumer Services, the Chemical Industry Council of North Carolina, the N.C. Association of Hazardous Materials Responders, the SBI, each Regional Response Team, and three persons appointed jointly by the North Carolina Fire Chief's Association and the North Carolina State Firefighters Association.

The act allows the secretary to contract with units of local government to provide Regional Response Teams. To date, these have been provided by selected fire departments around North Carolina. The contracted fire departments provide training and storage facilities for personnel and equipment for the Regional Response Teams. The secretary may agree to provide funding for most aspects of the Regional Response Program, but he or she, for local responses not requiring technical-level capability or standby time, has no obligation to provide reimbursement for cleanup activities after the containment of a spill. Responders enjoy a fairly broad grant of immunity from civil liability pursuant to NCGS 166A-14(a) and 166A-24 as long as the response does not involve willful misconduct, gross negligence, or bad faith. Service with a Regional Response Team does not deprive any individual of any benefits to which he or she is entitled pursuant to the Workers' Compensation Act or any other benefit program, state or federal.

When authorized to respond in the course of its duties, a Regional Response Team may enter any public or private property, including the scene of the event to which it is responding and any adjoining properties deemed necessary by the team in order to manage the emergency.

Anyone who causes a release of HAZMAT which leads to the activation of a Regional Response Team is liable for the reasonable costs of the response to and necessary mitigation of the problem. The secretary invoices the person in question and then pursues collection of any unpaid costs in the superior court of the county where the incident occurred. The monies collected in this fashion, together with any other funds received in support of the establishment and operation of Regional Response Teams, are placed in the Hazardous Materials Emergency Response Fund.

A member of a volunteer emergency response organization who is called to serve by the governor in a declaration of a state of disaster or to serve on a Regional Response Team may take leave from his or her civilian employment without pay and may not be forced to exhaust his or her vacation time, accrued leave, or sick leave to serve in such a capacity. However, the member may, if he or she chooses, utilize any accrued leave or vacation time to receive pay. In order to be entitled to these protections, the request for services must be made in writing by the director of the DEM or the head of the applicable local emergency management agency and directed to the chief of the responder's department and the responder's employer. However, these provisions do not apply to responders who have been certified to the DEM by their local emergency management agencies as essential to the employer's own ongoing emergency or disaster relief activities.

# Federal HAZMAT Law

The basis of Federal HAZMAT law is found in three Congressional acts: the Comprehensive Environmental Response, Compensation, and Liability Act of 1980 (CERCLA), the Superfund Amendments and Reauthorization Act of 1986 (SARA) (42 USC 9601, and statutes following), and the Resource Conservation and Recovery Act (RCRA). These laws supersede any state rules or regulations which are not consistent with their requirements. All three laws are comprehensive, complex, and implemented by extensive regulations within the CFR. The HAZMAT response rules applicable to the fire service are found in Chapter 29, Sections 1910.120, and sections following, of the CFR. Subsection (q) of the regulations is of particular importance to FDs because it describes certain basic rules which all departments must follow.

All departments shall develop a HAZMAT emergency response plan [29 CFR 1910.120(q)(1)]. When the word *shall* appears in a legal context, it means that there is no option regarding the specified requirement. The party instructed to do or not do something is liable for the consequences of the failure to comply with the requirement. Therefore, the rule mandates that *all fire departments* establish HAZMAT emergency response plans. Subsection (q)(2) lists the *minimum requirements* for the plan. They include preplanning and coordination with outside agencies, personnel roles, chain of command, training, communication, recognition of and prevention of emergencies, safe distances and shelters, site security and control, evacuation routes and procedures, decontamination, EMS, alerting and response procedures, PPE and emergency equipment, critique and follow-up, and a provision allowing the incorporation of any state or local emergency response plans into the emergency response plan in order to avoid duplication. If local or state plans are available, the prudent department should examine them prior to preparing its own plan in order to avoid duplications and, hopefully, save work and time. This especially is important for VFDs because their members often work full-time elsewhere and the preparation of a response plan is time-consuming.

Subsection (q)(3)(i) requires that a site-specific Incident Command System or Unified Command System (ICS/UCS) be utilized at each response and that all activities at the scene be coordinated through the person in charge of the ICS/UCS, assisted by the senior official of each responding unit. The rule defines a "senior official" as the person with highest seniority on the site with operational control. Typically, the officer in charge of the first-in unit passes command responsibility up the chain of command as senior officials with higher seniority arrive on the scene. The officer in charge of the first-in unit has complete responsibility for the scene until someone his or her senior arrives to assume command. The person in charge, or Incident Commander (IC), is required, to the extent possible, to identify the HAZMAT or conditions present, conduct a site analysis, and identify engineering controls, HAZMAT handling procedures for the problem, and maximum exposure limits. The IC then must implement appropriate emergency operations, utilizing appropriate protective equipment for all personnel, together with decontamination procedures. The IC is required to limit the number of emergency response personnel participating in the response operations and stage additional response assets at an appropriate location. The IC is required to appoint an official to oversee all matters relating to safety during the response evolution.

There may be instances when skilled support personnel (for example, heavy equipment operators) are necessary during the response. The rules allow their use, but the personnel must be briefed thoroughly regarding the event and supplied with appropriate protective equipment before being put to work. If specialists skilled in working with particular substances are to be utilized, the specialists must receive annual training or demonstrate an acceptable level of competence annually before their services can be utilized.

Subsection (q)(6) deals with training and qualifications. Currently (2019), the rules recognize four basic levels of training and qualification for responders: Awareness, Operations, Technician, and Specialist. The entry-level qualification is Awareness. Awareness requires that the responder be able to recognize a HAZMAT problem, identify the substance, understand the need for scene control, and be able to summon additional assistance. Operations-qualified personnel are able to perform the functions of Awareness-level personnel, initiate evacuation procedures, and initiate containment procedures. Technicians and Specialists possess all the qualifications of the Awareness and Operations personnel, together with significantly broader cleanup capabilities.

The remaining Federal HAZMAT rules are complex, lengthy, and well beyond the scope of this volume. However, they must be consulted by any department when it is preparing its HAZMAT-response plans.

Put simply, HAZMAT is a very expensive game to play. Departments contemplating response capability other than Awareness should understand that expensive training and equipment will be required for responders and that these items should be reviewed carefully when planning for the department's future financial needs. The costs of HAZMAT training and equipment will be ongoing and will, very likely, increase at a rate in excess of that of ordinary inflation. Once a department elects to assume a particular level of response capability, the law requires that it provide services at that level and to do so competently.

# Chapter 13

# NFPA 1403

This chapter will discuss two important aspects of NFPA Standard 1403 and how they relate to North Carolina's firefighters. The first, and most obvious one, is training. Known variously as "Chasing the Dragon" or "Dancing with the Elephant," and probably a dozen other names, interior firefighting is undoubtedly challenging, exciting, fulfilling, insanely dangerous, and great fun for most of us who have engaged in that avocation. However, the supervisor (paid or volunteer) who sends an untrained person into a confrontation with a working fire, or who attempts to train personnel improperly, does so at the risk of substantial civil or possible criminal liability. Although a recommended and not a mandatory standard, NFPA 1403 (hereafter "1403"), which establishes a standard for live-fire training exercises, nonetheless reaches deeply into fire department operations in North Carolina and exists as a tool to both train personnel and measure their performance.

No NFPA standard becomes a law unless an authority having jurisdiction adopts it as such, however, because it is a nationally-recognized standard of performance, it is a set of rules which a court may try to use when evaluating the performance of a fire department. This is because when an accusation is made in court, that court must find an example of "proper" performance or behavior in order to evaluate the circumstances which led to the initial accusation of misbehavior. Because 1403 is a nationally-recognized standard of performance and has been adopted with a few modifications by our Fire Commission as a statewide standard of performance, it is one which all departments in North Carolina should be following when conducting live-burn training exercises. After all, the object of the exercise is a SAFE and academically meaningful experience for the student.

Live-fire training exists for two primary purposes: to qualify otherwise untrained personnel to fight fire and to retrain and/or upgrade the training of otherwise qualified personnel. The large safety component of 1403 is intended to protect both the student and the instructor. It should never be used as a recruiting tool (as has been done in North Carolina in the past). Suiting-up an untrained person and sending that person inside for a "taste" of firefighting, for instance, is certainly negligent and could approach criminal disregard for the safety of the untrained person participating in the exercise. By following 1403, the student should be well-prepared for the exercise itself as well as cognizant of the safety precautions and procedures to be followed should an emergency occur during the training evolution.

NFPA 1403 requires that a first-time (rookie) student successfully complete a series of prerequisite courses under NFPA 1001 at least at the Firefighter I level (Safety, Fire Behavior, Portable Extinguishers, PPE, Ladders, Hose-Appliances-Streams, Overhaul, Water Supply, Ventilation, and Forcible Entry). It should be noted, however, that this list is subject to amendment by NFPA at any time and should be consulted regularly. This requirement does not mandate that these classes be completed in a "certification" context (through an authorized delivery agency), but there must be documentation in the training record of the participant/student that the classes and testing standards were 1001-compliant. The simplest method of achieving this, obviously, remains formal training through our community college system and OSFM. In this context, there is extensive documentation of the training and it's free of cost to any member of emergency services.

When a training transcript is unavailable, North Carolina requires that the chief of the student's department or his/her authorized designee sign a written statement attesting that the student has met the 1403 requirements to participate in the live-fire exercise. An officer or designee of a department who knowingly submits a false statement in this regard when asked to do so does so at his/her peril. Civil or possibly criminal liability could attach to the person signing such a statement related to training if the untrained person is killed or injured during the live-fire training evolution.

NFPA 1403 discusses training in a number of different environments and Chapter 4 discusses a very important one, the so-called "acquired structures." These are structures not constructed to be burned, but which are used for live-fire training. The fire service knows these as structures donated to fire departments to be used for live-fire training and then demolished. They make outstanding training tools, but they must be managed carefully.

In addition to the requirements of 1403, important North Carolina legal considerations are triggered by an acquired-structure live-burn training evolution, whether the evolution is run pursuant to 1403 or not. North Carolina imposes strict (and harsh) civil and criminal penalties upon someone who willfully damages another's real property without authority to do so. For example, burning the wrong building or burning it without the permission of the **actual** owner could lead to criminal prosecution and/or extensive civil liability. This problem often can be prevented by having a title examination performed by the department's attorney upon the property to be burned in order to be certain that the person offering the property to the department is in fact the owner and that there are no outstanding liens against the property (e.g., mortgages, tax liens, judgment liens, etc.). If liens are discovered, it is necessary to acquire the permission of the lienholders before burning the building. Otherwise, the department, those in charge of the exercise, and the property owner(s), may be held liable to the lienholder(s) for the amount(s) of the lien(s) or the value of the property destroyed. The legal fees for the title examination will be **substantially less** than the costs associated with the payment of the lien(s) or the reimbursement paid to someone whose building was burned in error.

Asbestos inspection and remediation is another 1403 rule with NC liability consequences if ignored. Because asbestos fibers become airborne and travel with wind and smoke during a fire, the stuff can have unexpected catastrophic health consequences for persons exposed to it. Both North Carolina and federal law classify asbestos as HAZMAT and require that it be handled by persons trained to do so and not casually spread around the community by an irresponsible live-burn exercise. A department and its personnel can be held liable for failure to address the asbestos mitigation requirements of 1403, regardless of whether or not 1403 is followed during the training evolution.

Air quality is another important, and bothersome, aspect of live-burn training. However, a little prior planning with the North Carolina air quality office having jurisdiction over the site of the exercise can prevent these problems. According to their publications, North Carolina's air quality offices are very willing to accommodate fire departments who wish to burn buildings if the departments will contact them ahead of time to make the necessary arrangements.

North Carolina law suggests rather strongly that when a department is running an "acquired structure"-type exercise, the department assumes absolute possession and control over the property from the time the exercise begins until the time when possession and control of the property is handed back to the owner. These events **must be** carefully documented **in writing.** During the training exercise, the department (and the instructors involved) are responsible for the safety of the site, and this includes the public, which will invariably appear to watch the proceedings and probably reappear after the exercise to "view the remains." By documenting the taking of control by those running the exercise as well as the return of control to the owner, it is possible to keep departments and instructors from being held accountable for an event which occurs at the site long after the exercise is finished, the departments involved have taken up, and everyone has returned to quarters. This unfortunate course of events has occurred in North Carolina in several instances and with tragic results.

Fortunately, 1403 and the live-burn training guidelines published by OSFM contain many of the forms necessary to manage the paperwork requirements of live-burn training. They should be followed religiously. Any instructor or department who has questions about these forms and their supporting documentation, or who wishes to modify them, should consult OSFM and an attorney before doing so. This is because much of this documentation is the result of someone else's mistakes and modification of the documents may cause you to repeat them.

The other aspect of 1403 is its indirect influence on a department's operations. Because 1403 requires a certain level of training before a rookie firefighter can engage in a live-fire training exercise under controlled conditions, it certainly is not illogical to assume that a court would expect a department to train its personnel at least through the NFPA 1001 courses specified in 1403 before sending them into an interior firefighting environment. Many departments in North Carolina, including the author's, have adopted this procedure. Most encourage their members to complete the entire Firefighter I/II 1001 training cycle (or the then-current NC equivalent), but will allow a member to participate in serious (including interior) firefighting only after completing the 1403-specified courses.

There will never be a substitute for experience, but the effectiveness of lessons learned thereby is always enhanced if there also exists a background of formal training in the same discipline. Lawyers, for instance, have known this for a long time. NFPA 1403 can be an effective tool to train personnel (and thereby enhance the safety and effectiveness of the department) as well as easily document the training of personnel, because, after all, the quality of the service we provide, the insurance premiums paid by our constituents, and the funds we seek through grant programs are all tied, to one degree or another, to our level of training.

## Chapter 14

# Junior Firefighters

Junior firefighters ("Juniors") are one of our most important resources as a means to recruit and train future firefighters. Juniors are minors and, therefore, function in a different legal context from adults. However, much of the law governing Junior programs is new, and it is changing because the idea of regulating these programs is new to North Carolina. This chapter is intended to outline the law governing junior firefighter programs and relates this law to the suggested rules offered as guidance for managing a junior firefighter program. As of January 2019, the law and rules for managing these programs have not been examined by the courts, so we in the fire service will be teaching ourselves how to properly utilize this body of law and rules. This chapter will discuss this context and offer a few suggestions from the perspective of a lawyer.

It is vital for anyone involved in a Junior program to understand that a Junior is a minor, **not an adult**, under North Carolina law. North Carolina law defines a minor as any unemancipated person under the age of 18 years. Here we have a clearly defined line which separates a minor from an adult, at least for purposes of establishing the point in time when a Junior can "convert" to an adult firefighter—his or her eighteenth (18th) birthday. In North Carolina, a minor, for instance, cannot enter a contract, convey real property, or waive his or her legal rights. This means that a minor cannot waive his or her right to sue someone else because of something that person did to the minor during the time the victim was a minor. The various members of the family of a minor can waive their rights to sue (assuming they are adults, of course) and can sign written consents for the minor to join the junior program, but they are not allowed to waive the rights of the minor to sue, even though they may be his or her parents. Except under very limited circumstances, one person cannot waive the legal rights of another. A minor who has been injured has from one to three years after his or her 18th birthday to file suit should he or she choose to do so (NCGS 1-15, 1-17, and statutes following). Should a lawsuit erupt, North Carolina's courts are required to consider the welfare of the minor as their primary concern (or "polar star") when making decisions. This inevitably places a heavy legal burden upon those accused of injuring the minor.

Many times, it is to the advantage of the Junior that he or she receive community college credit for classes successfully completed during the junior program. The North Carolina Community College System (NCCCS) requires that any minor enrolled in school obtain written consent from his or her school before enrolling in a community college class. In addition to assuring that the Junior receives credit for the class, this authorization also works as a means to facilitate a good working relationship between the secondary school attended by the minor and NCCCS. The NCCCS has a list of fire service-related courses for which it will not give a minor credit under any circumstances. It behooves any group administering a junior program to obtain this list and follow it religiously. Falsifying records to get a minor credit for these classes can cause catastrophic circumstances for the falsifier. These are official public records and falsifying them could lead to criminal charges of falsifying an official document, falsifying a transcript, false pretense, or even perjury if an oath or affirmation is involved in one of the falsified or altered documents.

The NCCCS as well would be expected to take some sort of action against a department or instructor who intentionally violated these policies. This could include removing the instructor involved from its list of approved instructors or denying the offending fire department access to certain services. The argument could be made that the NCCCS is open to all persons able to meet its admission criteria, but ignoring rules intended to protect students and assure the orderly operation of the community college system is not the function of the NCCCS. The member schools have an obligation to protect their own interests, including the welfare of their students.

The North Carolina Fire and Rescue Commission (NCFRC) has published a set of recommended rules for operating a junior firefighter program titled "Junior Member Standard." The primary idea behind the standard is protection of the Junior. This means very simply **no dangerous training environments for juniors**. The "Junior Member Standard" contains the same list of forbidden training classes for minors as do the NCCCS rules. In fact, it appears that the list was proposed to NCCCS by NCFRC and then adopted by both organizations. These rules are not legally binding upon anyone until a court says they are legally binding, otherwise, they are suggestions. However, they are the only set of uniform published rules available, and they're based on an examination of available law. They can be used to prevent injury to a minor or as a tool to accuse someone of not taking proper precautions to prevent injury to a minor, as the case may be. When someone is injured, the courts search for a standard of performance when trying to evaluate the conduct of someone accused of injuring another and the "Junior Member Standard" will be one of them when a lawsuit erupts which involves an injury to a Junior. North Carolina's courts have traditionally examined the facts surrounding an injury to a minor in great detail unless there are existing standards for conduct easily available for them to compare to the facts in the case before them. This standard makes it easy to stay within a zone of safe behavior when working with Juniors, but it is also a very straightforward set of rules a court can use to determine liability. When incorporated into a junior member training program, these rules, along with other rules and SOGs of the department serve two useful purposes. First, they provide a safe training environment, and second, they acclimatize the Juniors to the idea of SOGs and rules as a way to perform their duties successfully.

The NCFRC cannot penalize a department for not following the "Junior Member Standard," but following it certainly will not hurt the department during an administrative or insurance rating inspection. However, the NCFRC could use the standard as a basis to penalize one of its certified instructors who violated the standard and injured a Junior. Anyone doubting this should consult a NCFRC-certified live-burn instructor. As the creators and publishers of the "Junior Member Standard," members of the NCFRC also could be subpoenaed to testify against someone accused of causing injury to a Junior.

The creation of a standard for a training program involving minors has led to an examination of several aspects of North Carolina labor law. Initially, there was no provision in North Carolina labor law to accommodate the idea of a Junior program. So the General Assembly took the first step by amending NCGS 95-25.5 to allow Juniors to be treated as youth employees, subject, of course to all the usual restrictions relating to the employment of minors. It should be noted at this point that North Carolina labor law does not permit minors to be employed in dangerous environments. This amendment had the effect of opening the door to the possibility of Worker's Compensation and other insurance coverages for Juniors. However, as of January 2019, we do not have any rulings by a court or the North Carolina Industrial Commission that enrollment in a junior program automatically entitles the Junior to any coverages. This means that the department running the junior program has the duty and obligation to see to it that these coverages are put in place. This should be done **before** any Junior participates in any training. NCGS 95-25.5(b), when discussing youth (minor) employment, emphasizes the idea of keeping minors out of dangerous environments, and one can find this idea expressed directly or indirectly throughout youth employment rules. Because youth employment requirements are being applied to training Juniors, it's not particularly difficult to understand that these rules must be followed in the training process.

Worker's Compensation law is found in Chapter 97 of the North Carolina General Statutes. NCGS 97-2(2) allows minors (Juniors) to be included in Workmen's Compensation insurance coverages. However, as stated earlier, a department involved in the junior member program has what the law calls an affirmative duty to get the minor under the coverage. Because NCGS 97-7 applies Worker's Compensation rules to both state and local government, it would seem that the rules would apply to Juniors training with municipal and county departments as well as VFDs. NCGS 97-6 forbids an erstwhile employer from trying to get an employee (Junior) to sign an agreement giving away any coverage once it is made available. NCGS 97-10.1 forbids an injured employee (Junior) from suing his or her employer except under limited circumstances; however, the employee (Junior) can pursue third parties (e.g., instructors, community colleges, department members who caused the injuries) pursuant to NCGS 97-10.2.

Incorporating Juniors into training evolutions contains risk, so it is vital for every department or instructor conducting such training to document the status of any Junior who is participating. One method of doing so would be to identify each Junior who is participating and obtain a document from each Junior's department stating that the Junior is there with his or her department's approval for training approved by his or her department. The document should be signed by each Junior's chief or someone in the chain of command of the department who has authority to approve training activities. By so documenting the presence of each Junior, the department or instructor delivering the training should be protected from liability for an injury sustained by a Junior because the Junior should be covered by his or her department's Workmen's Compensation insurance.

From a lawyer's perspective, it is apparent that we in the fire service now have the tools to operate safe and productive junior member programs. We have a standard of performance from NCFRC and the door has been opened for Worker's Compensation and other insurance coverages, which must be put in place as soon as a Junior joins the department. However, we should follow the "Junior Member Standard" even though the Juniors may not be getting to live-fire and other activities until their 18th birthdays. Death or permanent injury sustained in the name of a "little fun" is both stupid and expensive. We have an absolute duty to these youngsters to show them the correct way to do things now so that when it becomes their turn to run things, they will continue the tradition of doing it correctly.

# Chapter 15

# Unlawful Burning

## Introduction

North Carolina has numerous General Statutes intended to deal with the burning of substances and those found in Article 15 of Chapter 14 deal with unlawful burning. In addition, some counties and municipalities have enacted local ordinances or, in some instances, local acts through the General Assembly, which deal with local burning issues. Any fire service official seeking enforcement of sanctions (penalties) against someone engaged in suspect or unlawful burning activities should review the ordinances of the AHJ before accusing the subject of the investigation. There may be excellent tools available at the local level which are not available in the General Statutes. This chapter will review only the unlawful burning activities presented in the General Statutes, in numerical order. The large number of counties and municipalities in North Carolina makes a review of local ordinances impractical.

## Arson (NCGS 14-58)

In some states, arson is a term used to describe many types of unlawful burnings. However, in North Carolina, it describes one specific crime. North Carolina uses the English common law definition of arson: the willful and malicious burning of the dwelling-house of another. Within that definition, North Carolina recognizes two degrees of arson. First-degree arson is the willful and malicious burning of another's house when the house is occupied. Second-degree arson is such burning of another's house when the house is unoccupied. The maximum penalty for first-degree arson, without any fatalities or serious injuries, is life imprisonment, and the maximum penalty for second-degree arson is forty years imprisonment.

Willful and malicious conduct is considered to be intentional conduct undertaken without excuse, justification, or bona fide claim of right. Ill will or animosity toward either the property or the owner is not a required element of the crime. Burning is defined as damage to the property. Simple charring is sufficient. A dwelling house is one which is inhabited (someone is living there) and can be any sort of abode, including manufactured homes, such as mobile homes and recreational trailers (NCGS 14-58.1 and NCGS 14-58-2), an apartment, a condominium or a similar occupancy. The burning of an apartment is arson, even when the fire is restricted to the apartment of the fire setter, because the *building* which is burned is occupied by other people. The same logic could be applied to a condominium or cooperative apartment. However, a townhouse might not fit the definition because townhouses, while attached with common walls are, in fact, separate buildings, owned individually. Close examination of the underlying documentation of the townhouse project is necessary in each instance in order to determine the type of dwelling involved. In order to show that the burned dwelling is that of another, it need only be shown that someone else lives there or that someone other than the fire setter owns or shares ownership in the property. The presence of someone within the burned building is the remaining element of first-degree arson. The person present need not be an owner or a tenant. The only requirement is that someone be present within the dwelling at the time of the fire.

## Public Buildings (NCGS 14-59) and Educational Institutions (NCGS 14-60)

It is a felony in North Carolina for a person to wantonly and willfully set fire to, burn, cause to be burned, aid, counsel, or procure the burning of the State Capitol, the Legislative or Justice Buildings, or *any* building owned by the state, a state agency, a state institution, a political subdivision of the state, or any other governmental or quasi-governmental entity. An example of a quasi-governmental entity is a water or sanitary district.

Likewise, it is a felony to wantonly and willfully set fire to, burn, cause to be burned, aid, counsel, or procure the burning of a schoolhouse or other building owned by a public school, private school, college, or other educational institution. The language of the statute is broad, including within its boundaries any type of building owned by an educational institution within North Carolina. Pranksters, especially those at the college level, would do well to keep this in mind. Setting a small fire in a dormitory or classroom could have unexpected and severe legal consequences for the fire setter.

## Fire Stations, Rescue Squads, Bridges, Certain Houses (NCGS 14-61) and Other Buildings (NCGS 14-62)

The wanton and willful setting fire to, burning, causing to be burned, or aiding, counseling, or procuring the burning of a bridge, fire station, rescue squad station, or house owned by an incorporated company or unincorporated association and used in the business of the corporation or association is a felony. Similar participation in the burning of uninhabited houses, stables, coach houses, outhouses, warehouses, offices, shops, mills, barns, granaries, or other buildings, structures, or erections used or intended to be used in any trade or manufacturing process or business whether in the possession of the fire setter, or not, is a felony. These statutes do not require an intent to act, as in a conspiracy; they merely require that damage occur by fire. In addition, the statutes do not indicate that the severity of these offenses is to be regarded as any less than that of arson. NCGS 14-62 was enacted to include many different types of structures. For example, the term *outhouse* has been interpreted to include other types of buildings in addition to the proverbial privy. In other words, you do not need a toilet (or bench and hole) to meet the statutory definition of *outhouse*. The statute is typical of those created to include large numbers of situations not otherwise covered.

## Buildings Under Construction (NCGS 14-62.1)

Whether a building is under construction or completed, to burn one is a felony. This statute is intended to penalize those who would burn, cause to be burned, or aid, counsel, or procure the burning of a building under construction. The statute specifically refers to buildings intended to be used as dwellings or in business or manufacturing, and then includes the phrase "or otherwise" in the same context, suggesting, at the very least, that the General Assembly had more in mind than just residential and commercial structures when it wrote the law. It does not matter whether the burned building belongs to the person causing the fire or to someone else. The reach of the statute appears to be quite broad.

## Churches and Religious Buildings (NCGS 14-62.2)

Churches, chapels, and meetinghouses used for religious purposes enjoy a higher level of statutory protection than some other nonresidential buildings. It is a higher-level felony to burn, set fire to, cause to be burned, or counsel or procure the burning of one of these structures. Such activity is also a federal crime. State and federal authorities have concurrent jurisdiction over such a crime. This means that, among other things, investigators from both jurisdictions have authority to act and that *both* court systems may prosecute an offender. This can result in an accused being tried, convicted, and sentenced in both state and federal courts.

## Vessels (NCGS 14-63)

It is a felony in North Carolina to burn, cause to be burned, or aid, counsel, or procure the burning of a boat, barge, ferry, or float in North Carolina. Again, there may be instances when state and federal authorities have concurrent jurisdiction. This depends upon the type of vessel burned, or the location of the vessel at the time of the fire. These jurisdictional determinations are fact-specific and will not be discussed in detail here. It is important for the investigator to remember that the statute exists and that the penalty upon conviction is that reserved for a felony.

## Tobacco Houses and Gin Houses (NCGS 14-64)

Certain agricultural buildings enjoy a unique statutory status. Buildings used to process tobacco and cotton (gin houses) are protected by this statute. To burn, cause to be burned or aid, counsel, or procure the burning of one of them is a felony. The statute makes no distinction between abandoned buildings and buildings currently in use. The old, seemingly abandoned tobacco barn in a field is covered by this statute, as well as the cotton processing plant which may have ceased operation. Once again, a prudent investigator would be well served to research the statutes before taking his or her findings to law enforcement authority. There may exist a prosecutorial tool available of which the district attorney is not aware.

## Fraudulent Burning of Dwelling Houses (NCGS 14-65)

Any person who occupies or owns a dwelling, or a structure intended to be used as a dwelling, and who wantonly and willfully or for a fraudulent purpose causes, aids, counsels, or procures the burning of the dwelling commits a felony. Typically, this crime involves an attempt to obtain financial gain, usually under an insurance policy, by damaging or destroying a dwelling. The interests of an investigator would be well served if, during the investigation, an exploration of the financial backgrounds of the owner(s)/occupant(s) were to be conducted. The intentional burning of a dwelling owned or occupied by the causer of the fire, without regard for the consequences to others as a result of the fire, constitutes a violation of this statute as well. Such crimes often involve mental depression, frustration, or a perceived idea that the fire will deny something to someone—for example, an estranged spouse.

## Burning Personal Property (NCGS 14-66)

Personal property is property other than land and buildings permanently attached to land. A manufactured home, for instance, if registered at the Division of Motor Vehicles, is personal property even though it is attached to land. Other examples of personal property are clothing, money, tools, sporting goods, furniture, books, toys, crops, farm equipment, vehicles, and almost anything else which is not permanently attached to land. Violation of this statute is a felony. To be convicted, it must be shown that the person accused intentionally or irresponsibly caused goods, wares, merchandise, chattels, or other personal property of any kind to be burned with the intent to prejudice the rights of an insurer, a creditor, the owner of the property, or anyone else. The property can be owned by the causer of the fire or anyone else. The legal relationship between ownership of the personal property and the person or entity whose rights have been prejudiced by the fire will be examined by the court at trial and established at that time.

## Other Buildings (NCGS 14-67.1)

Occasionally, law enforcement authorities may encounter situations which appear to be violations of the law, but they cannot find a specific statute to cover the situation. In the area of unlawful burning, the General Assembly has sought to deal with one of these situations. It enacted NCGS 14-67.1 to cover the burning of any type of building not described elsewhere in the General Statutes. The person who counsels the burning of, aids in the burning of, causes the burning of, or burns a building not otherwise described in the General Statutes commits a felony. This is a "catchall" statute to try to clean up any remaining "loose ends." Actual damage to the structure is not necessary under this statute. The courts have held that merely attempting to burn is sufficient for a conviction.

## Burning Caused During Commission of Another Felony (NCGS 14-67.2)

Effective Devember 1, 2018, it now is a Class D felony in North Carolina to damage in the value of $10,000 or more by fire or explosion any dwelling, structure, building, or conveyance or to in any way cause to be damaged any dwelling, structure, building, or conveyance in the value of $10,000 or more by fire or explosion during the commission of a felony unless the person's conduct is covered under some other law which provides for a greater level of punishment.

## Miscellaneous Misbehavior

It is a misdemeanor in North Carolina for the owner or occupant of a building or other premises to refuse to obey the lawful orders of a fire chief, code enforcement official, or the Commissioner of Insurance relating to features, facilities, or installations of buildings. The penalties are small—a fine of $10 to $50 per violation, per day.

Intentional failure by a municipal officer to investigate an incendiary fire is a misdemeanor, carrying a fine of an amount ranging from $25 to $200 per violation (NCGS 14-69). This statute should be kept in mind by anyone charged by a municipality with the responsibility for investigating incendiary activities. If an investigator becomes uncomfortable with a situation and desires to step aside, he or she is standing upon solid ground if he or she passes the investigation to other appropriate investigative authorities after notifying the municipality. This condition is especially important in small communities where the potential of a conflict of interest is high.

## Explosive or Incendiary Devices

Intentionally injuring a person with an explosive or incendiary device is a major felony in North Carolina. In at least one instance, the courts have said that throwing gasoline on someone with the intent of igniting it is a violation of this statute. It is a slightly lesser degree of a felony to damage property with such a device. Damaging a house of worship or a public building is a felony. This crime is categorized as injuring private property (NCGS 14-49). These offenses include in their definitions overtly committing the act as well as aiding, counseling, or procuring such crimes. If the damaged property is occupied at the time of the crime, the penalties can be even more severe (NCGS 14-49.1).

North Carolina defines an explosive or incendiary device as nitroglycerine, dynamite, gunpowder, any other high explosive, any other type of explosive or incendiary device, formulation, or compound; any substance capable of being used for such purposes; and any other similar compound or device when there exists some probability that the compound or device may be used to injure a person or property (NCGS 14-50.1). Federal statutes covering these types of crimes can lead to penalties being imposed in addition to any imposed by state courts. Whether or not a perpetrator will be subjected to both federal and state court jurisdiction is a case-by-case matter. The Oklahoma City event is an example of this type of concurrent jurisdiction. The identified perpetrators were first subjected to the federal courts, and, subsequently (November 2004), one of them was tried and convicted in Oklahoma's courts for the same offense.

## Bomb Threats (NCGS 14-69.1 and Statutes Following)

While not always regarded as a primary duty of a fire department, responding to a threat involving a weapon of mass destruction (WMD) has become one of the fire service's regular duties. Occasionally, the responding department becomes directly involved in the investigation of the incident, as the author discovered some years ago. An attempt was made to rob the bank in Newport, North Carolina, with the use of a false explosive device. Several agencies became involved before the investigation was completed, including investigators from the author's department. The perpetrators were caught and punished, but the most important lesson derived from the entire process was that of interagency cooperation. Each agency contributed something worthwhile to the effort, and the resulting assembly of information paid off handsomely.

It is a felony in North Carolina to make a report of a WMD if the reporter knows or has reason to know that what is being reported is false (NCGS 14-69.1). In addition to criminal penalties, the reporter, or perpetrator, of the false report can be required to compensate the victims of the report for financial loss resulting from the crime. A repeat offense is a felony of a higher degree. Communication of the threat by computer is also a violation of the statute (NCGS 14-668.1(e)).

It is a felony in North Carolina to construct a false device and display it in a fashion so as to suggest a threat (NCGS 14-69.2). A repeat offense is a higher degree of felony, with a more severe punishment. Also, as in the preceding statute, the perpetrator of the threat can be held accountable financially for the consequences of his or her actions.

In the 2003 session of the General Assembly, NCGS 14-69.3 was added to the General Statutes, effective December 1, 2003. It is a major felony in North Carolina if a firefighter or EMS worker is injured seriously in the course of responding to the previously described types of unlawful burning or false WMD activities. The statute does not define a serious injury. Presumably, this would be a matter to be decided during the trial.

## Woods and Brush Fires

Sometimes state and federal courts will have concurrent jurisdiction over woods or brush fires. Usually, this occurs when a fire damages federal lands which are also North Carolina game lands under the supervision of the Wildlife Resources Commission. This section will discuss only the North Carolina statutes involving woods and brush fires.

Intentionally setting fire to one's own woodlands without first notifying those owning or supervising the adjoining properties and without taking reasonable steps to control the fire is a misdemeanor if the fire damages the property of others. Intentionally setting fire to the woodlands of another with the intent to injure the property is a felony. Woodlands are defined as forest areas (both timbered and cutover) and second-growth areas which previously have been cultivated. This is a very general definition, obviously intended by the General Assembly to include as many different types of woods and brush as possible. In addition, this statute allows the accused to be sued for civil damages by the victims of the burning. It also provides for a reward of $500 to be paid to anyone who furnishes information to appropriate authorities which is sufficient to convict someone of a violation of the statute (NCGS 14-136).

Setting fire to any woods, lands, or fields of any sort, which are under the supervision of the Department of Environment and Natural Resources as part of the department's fire-control activities, whether the fire results from intentional or unintentional behavior, is a misdemeanor. The only exception to this regulation occurs when landowners fire their own open, non-wooded lands or fields in connection with farming or construction activities and confine the fire to their own premises at their sole expense. The courts have held that simply negligently causing the burning of lands under the supervision of the State Forest Service is sufficient to convict someone of a violation of the statute in the absence of any evidence of extenuating circumstances (NCGS 14-137).

Failing to adequately extinguish a woods or brush fire of any type which one has initiated is a misdemeanor (NCGS 14-138.1). This means that even though the fire itself is a lawful one, those who start it are responsible for making certain that the fire is fully extinguished. Under NCGS 14-140.1, it is a misdemeanor to fail to maintain a careful watch over a fire which has the potential to damage or destroy property. Any such fire which escapes its intended boundaries constitutes prima facie evidence of a violation of the statute. In essence, one planning to burn something must be sure the fire is well supervised.

Burning crops, pastures, or feeds (provender) can be either a misdemeanor or a felony. If the damage is valued at $2,000 or less, the crime is a misdemeanor. If the value of the loss exceeds $2,000, the crime is a felony (NCGS 14-141).

A side effect of any unlawful burning activity involves the by-products of the combustion. These are pollutants and are subject to regulation at state and federal levels. A troublesome repeat offender can be dealt a heavy financial blow by environmental authorities. Most of these offenders do not realize that the federal authorities, for instance, assess the fine and then place the burden on the offender to go to court to try to escape it. That fine has a life of at least ten, and possibly twenty years, if left uncollected, and it can be collected by the seizure and sale of the property of the offender. These regulations are useful tools in circumstances when other statutes do not provide clear guidance regarding burning control enforcement. There also are local ordinances or local acts scattered across the state which deal with burning. Many are concerned with the by-products of the combustion and should always be consulted when dealing with apparent burning violations.

## Defacing Flags (14-381)

This statute forbids the defacing of the flags of the United States of America and North Carolina, including mutilation (which arguably includes burning). While burning a flag has been held by the Supreme Court to be "protected speech" within the meaning of the First Amendment, the burning of personal property is a crime under both this statute and NCGS 14-66 described above. It would be interesting to see how a North Carolina court might deal with this issue.

# Chapter 16

# Liability

## Overview

*Black's Law Dictionary* offers several succinct definitions of liability based upon cases from around the United States, which, when read together, can give the reader a "feel" for the general concept. Some of them are

a. an obligation one is bound in law or justice to perform;
b. [the] condition of being actually or potentially subject to an obligation;
c. [the] condition of being responsible for a possible or actual loss, penalty, evil, expense, or burden;
d. legal responsibility;
e. the state of being bound or obligated in law or justice to do, pay, or make good something; and
f. the state of one who is bound in law and justice to do something which may be enforced by [legal] action.

North Carolina recognizes two basic categories of liability—criminal and civil. Criminal liability is accountability by someone for the consequences of the violation of a statute or other law which requires someone to do something or refrain from doing something. The courts sometimes refer to violations of statutes or other laws as public wrongs, or injuries to the community in its social aggregate capacity which adversely affect public rights and which are pursued (prosecuted) by public authority (the government). In shorthand, they are what we call crimes—for example: murder, arson, rape, larceny, embezzlement, breaking and entering, trespassing, driving while impaired, reckless driving, and so forth. If one is found liable for a violation of one of these (a verdict or plea of guilty), the penalties will involve imprisonment or a fine or both. To be held accountable, one must be found to have violated a specific crime, and the findings must be beyond a reasonable doubt. The imposition of a penalty must be performed utilizing a very strict set of rules, known as uniform sentencing, in order to assure that the due process and equal protection rights of the convicted criminal are not violated. Most of these crimes are found in Chapters 14 and 20 of the General Statutes. However, some chapters of the General Statutes (for example, those dealing with governmental finance) also contain specific criminal penalties for other types of conduct. The actual procedures for the enforcement of criminal laws in North Carolina are found in Chapter 15A of the General Statutes. This act establishes a uniform set of rules and procedures which must be followed *statewide* by all officials who are managing criminal cases.

Civil liability, on the other hand, is a private wrong, not directly involving the commission of a crime, and is pursued (prosecuted) by an individual through a process known as a civil lawsuit, or litigation. Imprisonment is not permitted in civil cases, except in very unusual circumstances, when, for example, someone behaves in a fashion which a court determines to be in contempt of court. Typically, this behavior involves disrespect of a presiding judge or other judicial official or intentional failure to obey an order of a court. In a civil case, the party found to be liable is ordered to do something, refrain from doing something, or pay money to the winning party. There are very few specific statutes which impose civil liability. Most situations resulting in civil liability arise from earlier cases in which the courts ruled that when a particular event or set of events occurred, the person or persons responsible for the event or events could be held accountable financially for the consequences of their actions.

The rules for managing civil cases are known as the North Carolina Rules of Civil Procedure. They are complex and should not be taken lightly. Most law students spend at least two semesters studying them. They are designed to assure, insofar as is possible, that all sides in the dispute are afforded a "level playing field" during the course of the case. For instance, each side is required to disclose to the other side and allow the other side to review, prior to trial, all evidence it intends to use.

A civil lawsuit is a request by one party (called the *plaintiff*) that the other party (called the *defendant*) do something, stop doing something, and/or pay the plaintiff compensation for a loss that was caused by the defendant. The lawsuit begins with the filing of a legal document, called the "complaint," in the office of the Clerk of Superior Court in the appropriate county. The papers are then served upon the defendant, using the Rules of Civil Procedure. After the papers have been served, the defendant files a response to the complaint (called the "answer"). When the answer has been filed, the next step is a lengthy (and very expensive) process called "discovery," during which the parties to the lawsuit (plaintiff and defendant) exchange information, try to negotiate a settlement of the dispute, and usually engage in arbitration or mediation outside court to try to resolve their differences. If this process is unsuccessful, the case is scheduled for trial.

A civil trial involves two activities. One is the "finding of the facts." This is the taking of testimony and the presentation of other evidence. The plaintiff presents its case first, and then the defendant presents its version of the case, including reasons why the plaintiff should not be allowed to be successful. These reasons are called *defenses*. The whole procedure is completed in front of the "finder of fact." The finder of fact is ordinarily a jury, but in certain circumstances, with the agreement of both plaintiff and defendant, it can be the judge alone. The case is often tried before a judge when the facts and/or law of the case are complex.

The other activity is the application of the law. The judge is responsible for seeing to it that proper legal principles are followed throughout the trial and that the law is explained adequately to the jury before it begins its deliberations. The process of explaining the law to the jury is called "charging the jury." When both sides have presented all their evidence to the jury and the jury has been "charged" by the presiding judge, the jury is allowed to enter secret deliberations to attempt to reach a verdict. When the trial is completed, either side in a civil case can appeal the result. Appealing a civil case is very time-consuming for lawyers and, consequently, *very expensive* for the clients. North Carolina law does not allow the losing party to pay the winning party's legal costs except in very limited circumstances, so the very expense of the process itself becomes an added stimulus for the parties in the dispute to settle their differences before trial.

Most fire departments encounter situations involving civil liability in two contexts—injury to a person or property and contract disputes. A contract is an agreement among the parties, or participants, to accomplish something. Liability becomes a consideration when one participant in the contract is accused of not following the terms and conditions of the agreement. This is known as a "breach of contract." The rules for deciding whether or not a breach of a contract has occurred are complex and generally very fact-specific, based upon the wording of the contract. Even the basic rules regarding contract disputes are beyond the scope of this discussion.

As a general rule, VFDs manage contracts in much the same fashion as any business entity in the private sector, with the possible exception of a contract for services made with a political subdivision of the state. Paid departments are ordinarily departments of political subdivisions of the state, and their contractual obligations are handled by the governing bodies and not the departments themselves. If a potential or actual contract dispute emerges, any department's interests would be well served by prompt consultation with legal counsel. Many such disputes can be resolved by negotiations managed by neutral parties (legal counsel or arbitrators) at substantially less expense than attempting to resolve the dispute in court.

How does injury liability work? In order to be held accountable in court for the consequences of one's actions, four conditions must exist. First, it must be shown that the defendant owed some sort of *duty* to the plaintiff to do something, refrain from doing something, or pay something. The duty can arise from a statute, a court case, a contract, good common sense, or any other activity or condition the court finds sufficient to create a duty. A second condition is a *breach of the duty*. This is a violation of the identified duty by the defendant. At this point, the plaintiff must show that whatever the defendant did was, in fact, not in compliance with the obligation to the plaintiff which was described in the duty. A third condition is a loss by the plaintiff which had or will have a dollar value. This loss is called *damages*. The burden is upon the plaintiff to show that there was or will be a dollar-value loss out there somewhere, as well as its amount. The fourth condition is a cause-and-effect relationship between the conduct of the defendant (breach of the duty) and the dollar-value loss to the plaintiff. This principle is known as *proximate cause*. The plaintiff must show that the actions of the defendant did, in fact, cause the financial loss or the conditions which will lead to financial loss by the plaintiff which are the basis of the lawsuit. *All four* conditions must exist in order for a defendant to be held accountable. If any one of them is missing or not proven at trial, the plaintiff will lose.

A common example of the injury-liability process is a traffic accident. For example, Joe is driving through town at fifty miles per hour and the speed limit in town is, according to the black-and-white sign by the side of the road, twenty miles per hour. Sam, the local insurance agent, begins to cross the street within the crosswalk, reasonably believing that Joe will slow down, as required by law, and miss him. Joe cannot stop in time and hits Sam, breaking Sam's leg. Joe had a duty to drive twenty miles per hour, imposed on him by the rules of the road and the black-and-white sign. By driving fifty miles per hour, he violated, or breached, that duty. Sam is groveling in the street, holding his broken leg, which definitely has a dollar value. Joe's high speed (breach of the duty) prevented him from stopping in time to avoid Sam, and, thus, proximately caused Sam's injury. All four conditions for a case for liability are present.

In a lawsuit, one lawyer, or group of lawyers, is attempting to assemble all four conditions for a case for liability while the other lawyer, or group of lawyers, is attacking one or more of the conditions in an attempt to force a collapse of the case or a verdict favorable to the defendant. This process often involves extensive legal hair-splitting, sometimes resulting in the alteration of a case which appeared to be simple and straightforward at the beginning into one which becomes significantly more complex and therefore expensive to pursue.

Much of the focus in these cases centers on the breach of the duty. In order for a defendant to be held accountable for a breach of a duty, the conduct of the defendant must be shown to have been negligent. North Carolina's courts have issued many definitions of negligence through the years, and one of the preferred ones is "the failure to exercise that degree of care for others' safety which a reasonably prudent man, under like circumstances, would exercise, and may consist of acts either of commission or omission." Basically, negligence means failing to use ordinary care. On the fire ground, for example, ordinary care is assuring that a water supply adequate for the evolution at hand is available.

Sometimes, violation of a statute can be negligence. This is called "negligence *per se*." Traffic cases are classic examples of this kind of negligence. A traffic statute is created as a safety rule to protect the public and must be followed. According to North Carolina law, to disregard the statute is negligence in and of itself. Violation of other statutes and regulations can likewise lead to *per se* negligence accusations. Conviction of almost any criminal offense can lead to a negligence accusation once the criminal matter is completed. The O.J. Simpson case is a classic example. Ordinarily, the civil case is not commenced until a criminal conviction has occurred. However, there is nothing under North Carolina law that prevents an "O.J."-type case from being filed at any time.

Sometimes, the negligent actions of more than one person can combine to cause injury to a person or property. The courts allow the combination of these negligent acts to establish a breach of duty. It does not matter that one of the parties involved may have been more or less negligent than the other.

When there is not a clear-cut answer regarding the conduct of one accused of a breach of duty, North Carolina's courts will use a three-step or three-prong test (TPT) to evaluate the conduct of the defendant. This test is especially useful when a fire department or firefighter, while performing prescribed duties, is accused of a breach of duty. Typically the sequence is this:

1. Was the defendant trained in the activity which is being examined?
2. Did the defendant perform in accordance with the training in question?
3. Were the training and resulting performance consistent with that of similar departments/firefighters in the same geographical area?

Ordinarily, if all three questions can be answered in the affirmative, the defendant will not be found negligent. Given the increasingly varied and unusual nature of fire service operations, it is vital for North Carolina's departments to understand the importance of this test.

## A Short Glossary

Several important terms are frequently used by lawyers and insurance companies when discussing injury, or tort, liability.

**Tort.** This term identifies an activity (an automobile accident, for instance) and the body of law which deals with injuries, and the causes of injuries, to persons or property not involving the breach of a contract. A tort is considered a private act to be resolved between individuals and not through the activities of the state, as in a criminal prosecution.

**Exposure.** This term describes the possibility of injury to a person or property and/or the extent of any monetary damages which might have to be paid if someone is found to be liable to another. During discussion of the case, a question frequently thrown back and forth by lawyers and insurers is, "What is our (or their) exposure?"

**Holding.** This term describes the ruling, or opinion, by a court which expresses its ideas regarding the meaning of a law. A court's opinion has the same force of law as any other written statement of the law. A typical use of the term is a statement which begins: "The court held that...."

**Defense.** This term describes a fact, legal principle, or other concept which can be used by an accused to escape liability for his or her actions.

## Duties

This section of the chapter will discuss some of the duties the fire service owes to those whom it serves. A breach of these duties is behavior which can lead, eventually, to civil liability. A generalized duty is imposed upon the North Carolina Fire Service by NCGS 58-82-1, which authorizes it to act. Since the statute authorizes the fire service to "do all acts reasonably necessary to extinguish fires and protect life and property from fire," the fire service, in effect, has been instructed to perform competently and effectively (reasonably) when called upon for assistance. The statute imposes an overall duty upon the fire service to meet the requirements of the three-prong test (TPT) previously described. This duty is especially important when a department engages in specialized activities—for example, HAZMAT operations or confined-space rescue. Once the department arrives on the scene and initiates actions to resolve the problem, this general duty to perform competently is triggered, in accordance with the TPT. Because a department has a duty to perform in accordance with its recognized levels of training and/or certification, a department which intentionally acts outside the scope of its training or certification can be held liable if persons or property become injured during the response. For example, a department whose members are certified in confined-space rescue will be expected by a court to perform competent confined-space rescue in accordance with current standards for that evolution. If an injury results from this activity (or from other activities requiring specific levels of training and/or certification), and it can be shown that the department failed to comply with the existing standards for the evolution involved, the department may be held accountable.

A related aspect of this duty is the one imposed upon municipal departments by NCGS 58-80-40, requiring that the municipality not be left unprotected. This statute should be kept in mind by chief officers and training officers when planning mutual aid agreements, responses into rural areas, and training exercises. For example, while taking the entire department out to a rural area to engage in a large water supply exercise may be excellent training, it may also violate the duty imposed upon the department by this statute to keep the municipality protected.

## Constitutional Rights

Departments owe a duty to respect the constitutional rights of their members and also those whom they serve. Occasionally, a lawsuit is filed in which a violation of a constitutional right is alleged. The legal basis for the accusation of a violation of constitutional rights used most frequently is a section of the United States Code, 42 USC 1983,the Civil Rights Act. These lawsuits are sometimes referred to as "1983 Cases."

Two constitutional amendments are popular targets when these types of cases involve fire departments—the First Amendment and the Fifth Amendment. The First Amendment, among other things, forbids the establishment of any religion by a governmental body. The courts have used this clause in the First Amendment to forbid any sort of religious activity by governments while performing their duties, including meetings. In North Carolina, it is not unusual for a governmental meeting to be opened with an invocation, typically led by a member of the organization holding the meeting. Courts have held that such an invocation is a form of governmentally approved religious activity—an act specifically forbidden by the First Amendment. The men who wrote our Constitution had observed the unpleasant side effects of religions established by governments and were determined to prevent such effects from occurring in America. Accordingly, they provided for complete freedom of religious expression and forbade any activities by a government which suggested or favored a religion. The courts subsequently examined the issue and determined that religious expression by governmental bodies walked dangerously close to the endorsement of a religion by a governmental body and that, therefore, religious activities by governmental bodies, *or their agencies*, amounted to a violation of the First Amendment. A privately incorporated VFD, not being a part of government, should not be subject to these rulings. However, a VFD which is a department of a municipality or other branch of local government is subject to them.

The author is aware of at least one instance of a public meeting in North Carolina where this First-Amendment issue was raised out of court: A member of a religion other than that under which the invocation was conducted requested that a member of the clergy for the religion of the member making the request also be allowed to conduct an invocation at the beginning of the meeting. The courts have held that if a government wants to allow one religious group to participate in its activities, it must allow all of them to participate. The governmental body conducting the meeting saw "the handwriting on the wall" and abandoned any further efforts at invocations. This issue has also been enforced by the courts in favor of those who do not follow religious practices—for example, atheists and agnostics. Courts have held that atheists' and agnostics' religious (or nonreligious) rights of expression, which are absolutely protected by the First Amendment, cannot be violated by governmental bodies requiring them to participate in religious activities. The argument has been made more than once that if someone doesn't agree with a religious activity in a meeting, that person is free to leave. The courts have said that to offer a person that option is, in effect, to deny him or her access to the meeting—another violation of First Amendment rights.

Presently, the most interesting case of this nature appeared to be the one involving the phrase "under God," which had been inserted into the Pledge of Allegiance by Congress in the 1950s. Filed on behalf of atheists and agnostics, this constitutional challenge was rejected by the United States Supreme Court in June 2004. (Its decision was based on a technicality, not on the lawfulness of the phrase "under God.") Among other arguments, those who pursued this case argued that the challenged phrase allowed a religious belief to be utilized in a governmental activity—the pledge itself. Another First Amendment case decided by the United States Supreme Court, involved the author's undergraduate college, the Virginia Military Institute (VMI). Some cadets sued, alleging that because VMI was a state-supported school, the invocation at mealtime in the cadets' dining facility amounted to a governmentally endorsed religious activity. (During the author's four years at VMI, both invocations at meals and attendance of services on one's Sabbath were required aspects of cadet life, and no one offered any particular objections other than those one would ordinarily expect from college students with very little free time available.) The Supreme Court agreed with the plaintiffs and forbade the practice as a violation of the First Amendment.

Both state and federal courts have held that any nonprofit organization (a VFD, for instance) or governmental agency must afford its members or employees due process of law under the Fifth Amendment when conducting disciplinary actions, including dismissal from the organization. For paid employees, this becomes a rather straightforward, if somewhat lengthy, process if proper procedures are followed under the State Personnel Act and or the department's personnel policies. The rules for VFDs are not as clearly defined. Due process requires that the person being disciplined be afforded an opportunity to confront his or her accuser(s), present his or her version of the events in question, and have the decision made by an impartial decision maker. In at least one instance, the North Carolina courts have held that a member of a nonprofit organization can be ejected only by the same means by which that person was inducted into membership.

Occasionally, circumstances occur on a fire ground which cause members of the departments involved to become suspicious of someone who is present. Ordinarily, they may desire to detain that person for further investigation, utilizing a so-called citizen's arrest. North Carolina law gives a specific definition of "arrest" and permits one to be performed only by a sworn law enforcement officer acting within the scope of his or her authority. NCGS 15A-404 forbids a private citizen (including a firefighter) from arresting another person unless the person performing the arrest is working under the direct instructions and supervision of a sworn law enforcement officer, who, in turn, is acting within the scope of his or her authority (NCGS 15A-405). However, a private citizen (such as a firefighter) may *detain* another person if certain conditions exist at the time of the detention. Specifically, a firefighter may detain another person if the firefighter has probable cause to believe that the detainee has, in the presence of the firefighter:

a. committed a felony;
b. committed a breach of the peace (a general term used to describe almost any variety of misbehavior);

c. committed a crime involving physical injury to another person;
d. committed a crime involving theft or destruction of property.

If the decision to detain someone is made, the method of detention must be reasonable. Unfortunately, the definition of a "reasonable means of detention" is left to the presiding judge, which means that anyone seeking to detain another must be careful about the means to be employed. If someone is detained, law enforcement authorities must be notified immediately and the detainee must be surrendered to law enforcement authority as soon as possible. Notifying law enforcement by radio or telephone is useful at this point because the tapes at the dispatch center will provide neutral documentation of the surrender and detention activities.

## Mutual Aid

A department which responds to a mutual aid call owes the same duties to the public it would owe if it were the primary responder. However, the department requesting the assistance may encounter a problematic situation. The fact that one or more additional departments are involved in the evolution does not diminish the duty owed by the requesting department to the public. North Carolina follows a common-law doctrine called *respondeat superior*, which states that an employer is legally responsible for the behavior of his or her employees as long as the employees are acting within the scope of their employment. North Carolina courts have held that a department responding to a mutual aid call is, in fact, an employee of the department requesting the assistance for the duration of the response; and therefore, the requesting department is liable for the actions of a responding department during its response to the request for assistance. Statistically, the most frequently encountered circumstances which expose the requesting and responding departments to liability are motor vehicle accidents involving the responding department. There is very little defense available to departments caught in this trap other than good insurance coverages and carefully worded mutual-aid agreements. The victim's lawyers usually sue every department with any potential liability, and the type of department, whether paid or volunteer, does not matter. The accusations will be pointed at all possible targets, and the plaintiff's lawyer will let the named defendants sort out the rest of it.

## Service Delivery

Perhaps the most difficult issues involving duties arise during the delivery of the services. The duty of the fire service is to protect life and property from fire—a deceptively simple definition of the duty, especially considering all the additional duties assumed by the fire service in the past twenty-five years or so. Yet each emergency is different and requires a department to take different actions to deal with the different problems presented, leaving the fire service with, seemingly, an infinite variety of possible responses to each call. This lack of a uniform method of response is, in one sense, a positive situation, because, other than those in the NFPA, there are not many exact standards of performance connected with the duty itself. This allows the fire service some flexibility in the manner of its responses. Unfortunately, the same absence of precise standards also opens the door for a plaintiff to initiate a claim against a department because there is nothing which says he cannot do so.

The three-prong test (TPT) requires that firefighters be trained, that the training meet a regional standard, at the very least, and that firefighters perform in accordance with the training. The NFPA is constantly recommending *nationwide* standards for service delivery which are classified as "minimal," and many of these standards are beyond the financial and physical capacities of many small departments. The NFPA 1500 standard for safety has been around a long time, as has the NFPA 1403 standard for participating in live-burn exercises and interior firefighting. More recently, two standards dealing directly with FD operations were adopted—1710 for paid departments and 1720 for VFDs. Both were adopted over vehement written and verbal protests from members of the fire service, including the author.

When applying the TPT, the courts are looking for a standard of performance against which to judge the performance of the defendant department. If nothing else is available, the court will turn to the NFPA standards governing the evolution(s) in question to decide if the requirements of the TPT have been met. Given the difficulties in meeting some of the NFPA standards, it is vital that North Carolina's departments provide some other guidance for the courts. The only way to do this is to write their own standards—called Standing Operating Guidelines (SOGs). By writing guidelines for performance which are suited to the needs and capabilities of individual departments, *training in accordance with these guidelines*, and then following them at the scene of a call, departments can take several giant steps away from liability for a claim against them based upon a failure to perform competently.

A word of warning is appropriate here. Both the State of North Carolina and the courts expect all departments to follow the 1403 standard closely (as modified by the OSFM). As of 2019, in the court system are several cases based upon violations of the 1403 standard where liability is being asserted, and the cases show excellent chances of success. If there is one NFPA standard which firefighters should follow religiously, whether paid or volunteer, it is North Carolina's version of NFPA 1403.

## Apparatus Operations

The rules of the road, the General Statutes, and good common sense, as discussed in Chapter 6, all impose upon the fire service a duty to operate motor vehicles safely and competently. The duty is owed both to the public and to fellow emergency responders. If an accident occurs and an apparatus operator or someone responding in a privately owned vehicle (POV) is convicted of a traffic violation, the convicted person is considered by the courts to have committed negligence *per se*, and the case for liability becomes much easier for the plaintiff. In the case of a POV, if it can be shown that the operator of the POV was responding to a call on behalf of a department, *respondeat superior* will likely be applied by the court to hold both the operator of the vehicle and the department liable for the consequences of the accident. When an apparatus responding to a mutual-aid call becomes involved in a motor vehicle accident, the operator of the apparatus (sometimes together with the person in command of the apparatus), the responding department, and the requesting department can all be held accountable. Invariably, these collisions result in severe damage to the POV and minor damage to the department vehicle.

The privilege to exceed the speed limit can only be exercised without endangering others on the road (and the crew on board). Responders are allowed to <u>ask</u> others to yield the right of way to them when they are operating under emergency circumstances with proper warnings, but they can do so only if that action can be made safely. Responders are not permitted to force the right-of-way issue. Appropriate use of warning devices is required, but if the POV will not yield the right of way, it must be left alone and should be reported to appropriate law enforcement authorities. A responder's POV is *not* an emergency vehicle under North Carolina law. When responding to a call, it must be operated in accordance with all rules of the road and other traffic laws, especially those relating to speed and the right of way. Lights and sirens alone do not an emergency vehicle make: it must be registered, licensed, and certified by the State of North Carolina, and owned by a duly authorized emergency services provider agency in order to be so classified.

In addition to examining the conduct of an emergency vehicle driver with regard to possible traffic law violations, courts now sometimes apply the three-prong test (TPT) to the actions of drivers. Questions regarding performance in accordance with training and the relationship of the training to similar training in the same geographical area point in one direction—formal training for all apparatus drivers, paid and volunteer. While North Carolina does not yet mandate formal training for emergency vehicle drivers, the day when this occurs may not be far away. The duty to drive responsibly remains and, by implication, places a duty on the department to train its drivers properly. A frequently recommended method of implementing the training plan is using a vehicle or apparatus response plan as part of the department's SOGs. Training a department's drivers in accordance with the plan and requiring them to follow the plan create a structure which delineates the duty to be performed and thus reduces the likelihood of a successful accusation of negligent breach of duty. Basic considerations for preparing a response plan recommended by some authorities in the field include the following:

1. Vehicle design, maintenance, and age;
2. Formal screening process for candidates;
3. Formal training of candidates;
4. Written duties and responsibilities for each driver-operator;
5. Intersection approach procedures and specific response speeds;
6. Nonemergency response procedures;
7. Backing procedures;
8. Hose loading procedures;
9. Periodic driving record checks on drivers;
10. Accident investigation procedures

Disciplinary procedures for serious violations of the driving SOG, when coupled with evidence that disciplinary action has been taken, will suggest to a court that the department is taking all reasonable steps necessary to insure that it is protecting the public against accidents. A driving SOG is a tool which can help negate accusations of negligence by the department. However, it also places a duty upon the department not just to create a plan, but to follow it.

# HAZMAT

The duties imposed upon the fire service relating to HAZMAT responses involve the interrelationship between North Carolina and federal law. Chapter 95 of the North Carolina General Statutes sets the general parameters for responses, including the conditions which require governmental supervision, while 29 CFR 1910.120 (federal law) describes the mechanisms and training required for responding. When examining the duties imposed upon the fire service, one must remember that state and federal law will frequently combine to define the duty. A common thread runs throughout these rules: the implied duty to perform competently when delivering the services, regardless of whether state or federal law is being applied.

One straightforward duty imposed under state law is that a department, especially the chief, protect and keep secure any industrial or trade secrets entrusted to the department as part of its hazardous-substance inventory information. Intentional release of the information outside the context of an emergency response is a felony (NCG 95-197); and a conviction in criminal court would establish negligence *per se*, throwing open the door to a successful civil lawsuit against the department. Unintentional release of the information, while possibly avoiding criminal charges, still leaves the door open for a lawsuit based upon the negligent release of the information.

Another, less clearly defined duty appears in NCGS 95-194(c). The statute mandates that any custodian of HAZMAT allow a fire department to enter its premises, after reasonable advance notice of the visit, to inspect the premises in order to conduct preplanning operations and check the accuracy of the Hazardous Substance List possessed by the department. This mandate imposes two duties upon the department. First, the department has a duty to take all reasonably necessary actions to obtain and update the HSL. While the statute does not appear to impose any penalties on the departments which fail to obtain HSLs from custodians, an accusation of failure to obtain such HSLs could become part of a civil negligence case against a department in a HAZMAT event. Second, a department may have a duty to conduct periodic inspections of custodians of HAZMAT located within its jurisdiction in order to preplan future operations and to update records. The preplanning issue has an interesting possible subduty as well. NCGS 95-194(e) allows the chief with jurisdiction to request a copy of the custodian's internal emergency response plan, which the custodian is required to furnish. Obviously, a prudent chief should obtain and include the custodian's plan as a part of the department's response plan. While this issue does not appear to have been raised openly in

North Carolina, as more and more manufacturing processes use chemicals, logic says it is but a matter of time before an alert trial lawyer raises it in a postevent lawsuit.

Federal HAZMAT rules are extensive and complex, but some of them found in portions of 29 CFR 1910.120 impose straightforward duties on firefighters and departments. As HAZMAT responders, departments are required by *federal* law to develop response plans. The wording of the rules does not suggest that developing a plan is optional. A department caught in a response without a plan may have committed negligence *per se.* If the department has a plan, it must train its members with the plan and use it in its responses if the department hopes to pass the TPT in court.

The TPT will likely be a part of any lawsuit involving a HAZMAT response, thereby placing a duty upon all departments to train and certify their members at some level of HAZMAT response and to perform accordingly. The plans and training themselves are not afforded the "local" flexibility of the North Carolina TPT because the specifications for training and planning are set by federal law and not state law. These activities must follow the federal rules. Because the CFR rules lay out the basics for each level of training and response, each level of response carries with it a duty to perform in accordance with the appropriate level of training at each event. Therefore, the department which exceeds its level of training in its response runs a significant risk of being sued successfully if a problem occurs during a response. In some instances this departmental failure to follow the CFR rules could even extend liability to an individual responder who operates, unsuccessfully, outside his or her level of individual HAZMAT certification.

A vitally important duty concerning all HAZMAT responses, given the dangerous nature of the activities under way, is to assure that the activities of the scene are conducted with as much regard for safety as possible. This duty for safe operations extends to both the responders and the public. In this regard, 29 CFR 1910.120 mandates that a safety officer be designated at all responses and that the safety officer be granted very broad authority to deal with safety issues. An IC who fails to designate himself, herself, or another person as safety officer does so at his or her peril.

In summary, departments have a duty under federal law to prepare response plans, to train their members both as responders and as participants in the plans themselves, to perform in accordance with their plans and training, and to do so as safely as possible.

# Occupational Safety and Health (OSH/OSHA)

Inherent in all OSH rules is a basic duty—to provide for a safe working environment for an employee. This duty is doubly important for the fire service because most, if not all, of its activities (including much of its training) are dangerous by their very nature. "The rules are the rules are the rules," so to speak. The duty imposed by the rules is an absolute one, with very little latitude for interpretation (not unlike the Building Code). The rules are to be followed very closely, with the only deviations being those approved by the OSHA.

Violations result in financial penalties assessed against the violator, accompanied by rather pointed instructions regarding the correction of the problem. The statute, fortunately, states that an OSH violation is not negligence *per se*. However, the confirmation of an OSH violation is admissible in court as evidence of negligence. Typically, the court admits such evidence when a worker is pursuing a "Woodson" claim or when an OSH violation was part of the proximate cause of an injury to a third party—for example, a bystander at the scene of a response.

The role of the Incident Commander (IC) or the commanding officer of a Unified Command System (UCS) in this regard is becoming more complex, especially when mutual aid is involved. Within the past five years, both North Carolina's courts and the Industrial Commission have begun examining the relationship between the primary (or general) contractor, the subcontractors, and OSH. In most respects, the relationship between a department with primary responsibility for a call and its mutual-aid responders is analogous to that of a primary contractor and his or her subcontractors. The courts and the Industrial Commission are saying that a primary/general contractor owes a duty under OSH to provide a safe work environment for the subcontractors. The extent of this duty into activities on the work site has yet to be defined clearly, but it does exist and primary/general contractors have been penalized for OSH violations involving subcontractors. As yet, there are no published cases relating to fire departments and mutual-aid situations, but it is likely that these rules will be applied to the fire service. This likelihood translates into the need for the IC to be sure that a properly trained and designated Incident Safety Officer is present *and enforcing safety rules* at all responses where OSH rules are applicable. For instance, in a situation where both paid and volunteer departments are present, the IC, as well as the supervisor of the paid personnel, are responsible for ensuring that the activities of the paid personnel are conducted in accordance with appropriate OSH standards. This situation should be a stimulus for all OSHA-exempt VFDs to plan and train in a manner as close to existing OSH standards as possible in order to simplify the duties of a volunteer IC at a scene where paid personnel are present.

An important part of the firefighter OSH standards is personnel accountability on the fire ground. It should be an important part of the operations of every VFD as well, because, although as yet not indicated by the courts, there likely exists an implied duty for volunteer ICs to keep proper account of their personnel present on the fire ground. This duty, of course, translates into a personnel accountability system. Recent events in North Carolina, including the sad incidents surrounding the deaths of two volunteer firefighters in Wayne County in November 1998 and subsequent injuries at live-burn exercises in other parts of the state, point increasingly in the direction of the duty of an IC to account for the personnel on the scene at all times. The litigation resulting from some of these events could lead to a court-created ruling regarding personnel accountability.

## Hiring

Both paid and volunteer departments owe an implied duty to the persons they serve to screen all applicants for employment or membership. The events at the World Trade Center have made this duty even more important. The level of screening should be determined by the nature of the duties intended for the applicant. Discriminatory screening can lead to an accusation of liability under the Civil Rights Act of 1964. Failing to screen adequately can result in liability if the matter which led to the questioning of the screening process was one which could have been prevented by adequate screening of the applicant. Successful lawsuits which held employers accountable for negligent screening of applicants have been filed in North Carolina and other jurisdictions.

Recently, as part of its response to the requirements of the Homeland Security Act, the General Assembly passed a law requiring that the FBI and SBI participate in the preemployment screening process for emergency responders, including background checks and fingerprint screening. The costs of such activities may not have been taken into account at the time, because, currently, the law is not being enforced, pending resolution of matters relating to costs and to coordination between state and federal authorities.

An important part of any screening process is the exchange of information between the potential employer and current and/or previous employers. Until 1997, this was a significant problem for some employers because of privacy considerations. In 1997, the General Assembly enacted NCGS 1-539.12, which provides for immunity from civil liability for employers who disclose certain information about current or former employees to prospective employers. The request for information must be made by a current employer, a prospective employer, or the employee, and the information which can be released is that relating to job history or job performance. A party who releases information which is false or which the releasing party reasonably should have known to be false can be liable for damages. According to the statute, "job performance" includes the suitability of the employee for re-employment, the employee's skills, abilities, and traits as they may relate to the job being sought, and, in the case of a former employee, the reason for the employee's separation from the employer being questioned. The statute applies to any employee, agent, or other representative of the employer being questioned who is authorized by the employer to release such information, as well as various types of employment services and job listing agencies. The availability of this information places a duty upon any employer or prospective employer to seek this information during the course of the employment process. Clearly, this is a duty for any paid department, and it well may be one for a VFD when one considers the possible ramifications of Homeland Security regulations.

Under the Fair Labor Standards Act, the employer has a duty not to discriminate in hiring personnel. All applicants must be given equal opportunities when applying for a job. However, this duty does not require that an employer hire someone obviously not qualified for the job as long as all applicants are screened using a uniform set of criteria. Because the FLSA does not apply to a privately incorporated VFD, the selection of volunteer members can be made by whatever means the department may choose. A VFD also has a hiring flexibility not available to paid departments, which must operate as public employers. Because it is a private, nonprofit corporation, a VFD may set many of its own criteria for employment in much the same fashion as any other private employer, and it possesses great flexibility in the method it chooses to select volunteer members. However, a VFD should be careful with its selection process in order to avoid unnecessary accusations of employment discrimination or discrimination in its membership selection process. Such nondiscriminatory practices could lead to performance-based qualification processes becoming part of both paid and volunteer departments.

Tests to determine physical qualifications for employment are overseen, at least indirectly, by the FLSA, the Civil Rights Act, the ADA, and North Carolina's Handicapped Person Protection Act. Between them and their supporting regulations, these laws impose a duty upon the fire service to conduct physical screening in a fair and uniform fashion. The body of law dealing with disabled persons also imposes a duty upon public agencies to provide physical facilities which, at least to some degree, accommodate the special needs of the disabled. The FLSA also imposes a duty to follow applicable wage and hour rules. Failure to follow these rules frequently leads to claims under FLSA regulations and/or lawsuits which have resulted in payment to employees by the offending employer of substantial sums in the forms of overtime pay, penalties, and interest.

## Firing

The acts of dismissing and disciplining a public-sector employee (paid firefighter) require the employer to follow constitutional due process of law. The employee is entitled to notice of the accusation, an opportunity to present his or her version of what occurred, and a review of the disciplinary decision by an impartial decision maker. Typically these requirements are addressed in the department's personnel policy. With paid departments, the procedures for dismissing and disciplining an employee will ordinarily involve a local personnel ordinance or the State Personnel Act. But because a VFD is legally a private employer and North Carolina is not a "right-to-work" state, the VFD may write its own disciplinary rules for paid members. The only duty imposed upon VFDs regarding their members appears to be one stated in a case involving a nonprofit real estate organization. That court ruled that a member of a nonprofit corporation must be removed from membership utilizing the same method which was used to select the person for membership.

## Health Insurance Portability and Privacy Act (HIPPA)

This remarkably complex piece of legislation contains at least one duty which has a direct impact upon the fire service—the protection of the privacy of the medical records of any patient served by the department. This duty applies to all providers, whether paid or volunteer. The law requires that any service provider possessing health information regarding a person take steps to assure the security (and thus privacy) of the information in its possession. Careless or unauthorized release of health information can have serious consequences under HIPPA. Both civil and criminal penalties are applicable. This body of rules, which took effect in April 2003, will affect those departments who perform any type of EMS and should be monitored carefully by all service providers. A detailed discussion of these rules is impractical because of their volume, scope, and frequency of amendment. One rule of thumb which has recently emerged requires that a service provider release only the minimum amount of information necessary for the purpose at hand. This rule should be applied within the context in which the decision to release is contemplated. An excellent way to develop a HIPPA privacy plan for a department is to do so in consultation with the provider hospital and legal counsel, utilizing model plans, drafts, and copies of the plans of other departments or hospitals. The emerging trend in North Carolina is to allow the provider hospital to develop the policy for the emergency services providers with which the hospital is affiliated. This trend allows for a uniform set of rules for all responders affiliated with the provider and saves legal expenses (a very important budgetary consideration for most departments, especially VFDs).

## Sexual Harassment

When American society recognized that preferential or discriminatory treatment or harassing behavior based upon gender was, in fact, a violation of the victim's civil rights, Congress and the courts acted to deal with the problem. The result has been a body of law generally referred to as "sexual harassment law." The fire service, like other areas of employment, now has a duty to its employees to prevent behavior which would be characterized as sexually harassing. This duty requires that all employers adopt a no-tolerance policy regarding such behavior, investigate any such complaints, and take immediate remedial action if the complaint is shown to be justified. Until June 1998, a complaining party had to show that some economic or physical harm had resulted from the behavior about which the complaint was made. In June 1998, the United States Supreme Court issued two opinions which expanded the definition of sexual harassment. The Court ruled that an employer could be held accountable for the consequences of a sexually hostile working environment even though no physical or economic harm had resulted from the behavior. In August 2001, the North Carolina Court of Appeals issued a quite different opinion in *Parik v. Eckerd Corp.*: The court held that boorish behavior did not amount to sexual harassment because it "did not affect a term, condition or benefit of ... employment." Notwithstanding this opinion, the fire service, like any other employer, still owes a duty to its employees to treat them with dignity and respect and keep the working environment as pleasant as is reasonably possible.

## Corporate (Nonprofit)

The duties imposed upon a corporation (a VFD) are, for the most part, those imposed by statutes (Chapter 55A) and those discussed in the foregoing portions of this chapter. The performance standards (duties) for directors are those set out in NCGS 55A-8-30. A director who acts in good faith, in the manner of an ordinary and prudent person under the existing circumstances, and in a manner which he reasonably believes to be in the best interests of the corporation has fulfilled his duties adequately. The statute also describes the types of information upon which a director can rely in making decisions. A director should always avoid conflicts of interest or situations which give the appearance of conflict of interest. Apparent, but not actual, conflicts of interest can lead to unnecessary discord within departments and criticism from its constituents, and therefore they should be avoided.

The actions of corporate officers are to be undertaken under standards which are much the same as those for directors—good faith, reasonableness, and the best interests of the corporation (NCGS 55A-8-42). Likewise, their actions should not involve conflicts of interest or the appearance thereof.

The foregoing discussion of duties is not all-inclusive. The nature of law makes a complete list of appropriate duties virtually impossible to compile. The ingenuity of trial lawyers and the ever-changing interpretations of the law by the courts work together to create a potentially limitless list of duties. For instance, who would have thought that a fast-food restaurant could be held accountable to a customer who was careless enough to spill hot coffee after purchasing and taking possession of it?

Practical jokes and skylarking in quarters are deep-seated traditions of the fire service and serve a useful function in stress release. However, any such activity should be undertaken in a fashion not likely to cause physical harm or mental stress which could be characterized as discriminatory or harassing.

## Defenses

Defenses are situations and laws which protect potential or actual defendants from liability. Defenses are available in both criminal and civil contexts. In a criminal context, paid departments and their members and VFDs and their members are treated similarly by the courts—if you break the law, you're accountable for your actions. Because of the highly sensitive nature of many inquiries regarding criminal behavior, the author recommends that questions regarding criminal behavior be directed to privately retained legal counsel if a specific activity is in question, and to a local district attorney if the question does not involve a specific activity which has occurred or is ongoing. However, North Carolina law recognizes distinctions between paid departments and VFDs in matters relating to defenses in a civil, or noncriminal, context. This portion of this chapter will examine defenses in the civil context. But it must be remembered that a criminal conviction is negligence *per se* and frequently creates a very strong case for civil liability.

## Sovereign Immunity

This legal concept is at the very heart of the liability defenses available to paid departments and their members. What follows is a brief discussion of how it functions.

According to the English common law, one is forbidden to bring a civil lawsuit against the Crown (the government). This restriction came to North Carolina with the early settlers and became an important part of North Carolina law. Today, it is often referred to as "qualified" or "governmental" immunity. For purposes of this discussion, it will be referred to by its older name, "sovereign immunity." The basic rule is that a unit of local government (municipality or county) and the State of North Carolina or one of its agencies can be sued only if one or more exceptions to sovereign immunity exist. Bringing a successful lawsuit against the state is very difficult. The cases brought against the state and the DOL as a result of the events in Hamlet in 1991 are excellent examples of this problem. State government immunity is beyond the scope of this volume, so this discussion will be limited to issues relating to local government.

Ordinarily, an accusation is made that some person or a group of persons affiliated with local government did something which injured someone, someone's property, or someone's constitutional rights. Because local government acts through its officers and employees, an attempt is then made to hold local government accountable for the specified actions by using the legal doctrine of *respondeat superior*, which states than an employer is liable for the actions of his or her employee if the actions were undertaken within the scope of his or her employment.

In a typical situation, the plaintiff will name the individuals believed to have caused the problem, together with the employer (a unit of local government), as defendants in the lawsuit. To establish a basic case for liability, the plaintiff must prove that three conditions exist:

1. That those accused of the negligent act(s) were employees or elected officials of the local government unit when the event(s) occurred.
2. That the activity complained about expressly was authorized.
3. That the activity complained about was either within the scope of employment (in furtherance of the business of the local government unit) or that the local government unit later authorized or ratified the activity.

If the plaintiff cannot establish any one of the three conditions, his or her case should be dismissed by the court. The unit of local government can win the case by proving successfully that one or more of the three prongs of the test listed above did not occur or does not exist.

Usually, the existence of the employment relationship can be established simply by requesting copies of employment documentation or contracts. Scope of employment often becomes a focal point of examination by the lawyers and the court. The employee must be shown to have been performing his or her duties, and an elected official must be shown to have been acting pursuant to his or her elected office. Sometimes, the nature of the activity can play a role as well. The courts recognize two acceptable classifications of activities for local government—governmental and proprietary. If an activity is found by the court to be proprietary, sovereign immunity rules are relaxed and it becomes easier to pursue the case. Some recognized governmental activities involve condemnation, fire protection, EMS, law enforcement, and public health. Some proprietary activities involve public utilities, public hospitals, airports, arenas, and convention centers. The distinction between these two classifications is that governmental activities are those either required by law or generally expected of local government, whereas proprietary ones are optional governmental activities and/or those that generate nontax revenue. There is at least one activity where the courts are all over the map with their opinions—operating and maintaining parks and playgrounds. At this time the answer apparently depends on which court is reviewing the case and what was being done at or with the park or playground at the time the incident occurred.

Even though a plaintiff may be able to establish the necessary elements of the foregoing three-prong test, he or she may still lose the case if there is insurance. A local government entity remains protected by sovereign immunity except to the extent it has liability insurance coverage. Two statutes, NCGS 160A-485 and 153A-435, allow such cases to go forward only if the local government entity has liability insurance; and if it has insurance, the plaintiff can sue only for the money in the insurance policy. For example, if Larsen Greasepalm sues the Town of East Armpit for damages resulting from misbehavior by a municipal employee and is awarded a judgment for $5 million and the town has $1 million in liability insurance coverage, Larsen gets only the $1 million. Recently, Mecklenburg County was sued successfully by an inmate from the county jail who had been injured. The judgment was for $49,500, but the inmate did not receive any money because the county's insurance policy had a deductible of $250,000 and, the court ruled that insurance coverage which would have allowed the plaintiff to be paid did not go into effect until the verdict exceeded $250,000.00. Thus, insurance did not exist for his judgment and sovereign immunity still applied.

In the late 1980s or early 1990s, a property owner in Atlantic Beach, N.C., converted an old beach house to apartments without building permits or inspections. The tenants moved in, and shortly thereafter a fire occurred in which a tenant died. The family of the deceased tenant sued the town for negligence in enforcing the required building code—that is, for failing to issue permits or conduct inspections. The town responded that it was immune from the lawsuit under sovereign immunity because it did not carry liability insurance. The attorneys for the plaintiff spent months searching for some form of insurance policy, were unable to find one, and had to dismiss the lawsuit.

Sometimes there exists something which looks like insurance but may not, in fact, be insurance. This is often referred to as a "risk pool." Ordinarily, one or more units of local government contribute money to a fund or pool whose contents are used to deal with liability claims made against any unit which has contributed to the fund or pool. Courts scrutinize these arrangements carefully on a case-by-case basis to determine if they are insurance plans or something else. If the arrangement is found to be an insurance plan, a plaintiff can pursue the money. However, if the arrangement is found not to be an insurance plan, sovereign immunity will apply to the case.

So what happens when the plaintiff names individuals as well as the unit of local government in the lawsuit? The Supreme Court of North Carolina says that liability determination will ordinarily be made based upon the language in the documents filed with the court. If the plaintiff is asking for the unit of local government to do or not to do something, the public official is protected by sovereign immunity. If the plaintiff is asking for monetary damages, the court must decide from the evidence before it whether the plaintiff is suing the individual or the government or both. If the plaintiff sues the individual, there may be times when the individual is protected by sovereign immunity.

The courts have recognized a difference between a *public official* and a *public employee*. A public official (or public officer) is sometimes protected from liability if working within the scope of his or her duties when the event complained about occurs. In *Willis v. Town of Beaufort et al.*, the North Carolina Court of Appeals held that a fire chief in the course of performing his or her duties is a public official. Unless a public official engages in activities for corrupt reasons or acts outside the scope of his or her authority, the official will generally be protected from personal liability.

The courts look at four factors when determining the status of someone in the public sector of employment:

1. Was the person's position created by legislation?
2. Does the position require an oath of office?
3. Was the person performing legally imposed public duties?
4. Was a certain amount of discretion required for the job?

There does not appear to be a set of required answers to these questions. However, cases suggest that the more affirmative answers that can be made, the better. The second question *must* be answered in the affirmative, however. The constitution of North Carolina requires that all public officials take oaths of office before performing their duties. This requirement is stated unequivocally in Article VI, Section 7 of North Carolina's constitution and in NCGS 160A-61 and 153A-26. Clearly, one "giant step" toward public officialdom and the immunities it carries is an oath of office.

There is no governmental immunity when a public official or employee uses his or her public office in a fashion intended to injure someone's person or property. The law calls this an intentional tort. Both the governmental official/employee and the unit of government (as employer) can be held accountable. In a case in Gaston County, a building inspector engaged in a "cat-and-mouse game" with a property owner who was trying to convert a single-family dwelling into apartments without the required permits or inspections. After some months of this activity, including an unsuccessful stop-work order, the inspector obtained a shaky administrative search warrant and announced to one or more witnesses that he intended to "teach the [property] owner a lesson." According to the lawsuit, the warrant, among other things, did not identify the apartments in the structure. The inspector then entered the structure, forced an entry into two of the apartments, inspected the structure and the two apartments, and submitted a list of code violations. On appeal of the resulting lawsuit, the Court of Appeals pointed out that there was no immunity because the evidence supported the plaintiff's claims that the inspector abused his authority by using the police power of his employer with the express intent to cause injury, to "teach the owner a lesson."

NCGS 160A-293 affords some specific immunity to a municipal fire department. It augments the immunities granted by sovereign immunity in its subsection (b). The statute states that no municipality or any officer or employee shall be held liable for the consequences of a failure to respond or a delay in responding to a call for fire protection outside the corporate limits. In subsection (c), it states that any employee of a municipal department shall retain all the authority and immunities available to himself, herself, and the municipality while working outside the municipal boundaries which are available while working within the municipal boundaries. However, in 1995, the Court of Appeals handed down a strange opinion in the case *Davis v. Messer* (119 N.C. App.44).

In this case, the Waynesville municipal fire department was accidentally dispatched to a fire in a rural fire district. The Waynesville department acknowledged the call by radio and initiated a response. Meantime, the dispatch center realized its mistake and alerted the proper department. During all of this confusion, the Waynesville chief realized that the department was leaving the municipality for another department's fire and recalled his personnel and equipment. The house burned down and the resulting lawsuit arrived in the Court of Appeals. The court found that Waynesville was, in fact, liable for the consequences of its fire department's failure to respond. In a convoluted opinion, the court found, among other things, that Waynesville was liable because its department's failure to respond was deliberate and that the words "failure or delay" in Subsection (b) of the statute were intended to mean unintentional failures or delays and not deliberate recalls. The court went on to state that if the department had waited until it crossed out of the municipal fire protection response boundary before turning around, Waynesville would not have been liable! Go figure. This case was subsequently reversed for other reasons by a 2001 case, *Willis v. Town of Beaufort*, but it remains on the books as an interpretation, however unusual, of the statute. The great lesson of this unfortunate mess appears to be that if a municipal department is dispatched to another department's fire without a mutual aid agreement covering the situation, it should either notify the dispatching authority of the mistake *and not respond*, or if the decision to respond is made, it should not turn back on its own initiative. It should let the dispatching authority or someone else with appropriate authority make the decision.

Ordinarily, sovereign immunity is a defense considered only in the context of municipal or county paid fire departments. However, two important cases applied it in the context of volunteer departments. One was the federal *Geiger* case from the Middle District of North Carolina; the other was *Luhmann v. Hoeing et al.*, a unanimous ruling which was handed down by the Supreme Court of North Carolina on 25 June 2004. *Luhmann* is a case to be examined carefully. Mr. Luhmann's leg was crushed while he, as a bystander, was standing between an engine and a tanker when the tanker operator (Mr. Hoenig) pulled away to refill without disconnecting the supply line from either apparatus. Mr. Luhmann sued the operator and his department (Cape Carteret Volunteer Fire and Rescue Department, Inc.) and won a $950,000 verdict at trial. On appeal, the North Carolina Court of Appeals held that NCGS 58-82-5 applied and that the fire department was not liable. One judge dissented, however, arguing that the privileges and immunities described in Chapter 69 applied and that, in fact, the department was protected by sovereign immunity, having the same privileges and immunities afforded to the agency of local government with which it had a contract for services (Carteret County). This would mean that the absolute release from liability afforded under NCGS 58-82-5 would not apply. The case was appealed to the Supreme Court of North Carolina.

The Supreme Court reversed the decision of the Court of Appeals and held that the department was, in fact, protected by the privileges and immunities of the county rather than NCGS 58-82-5. Because both the county and the department carried liability insurance, this meant that Mr. Luhmann could receive payment, in this case, from the department's insurance company. The court examined the day-to-day operations of the department and noted that the department contracts with Carteret County to provide services within a district, that Carteret County collects and disburses a fire tax within the district, that the tax revenues fund approximately 98 percent of the department's budget, and that the department has a number of paid employees. The court then held that the department was, in fact, a fire district within the meaning of NCGS 69-25.8 and that, therefore, pursuant to that statute, the department had the same immunities and authority as the county when delivering its services. This part of the ruling raises an interesting question regarding the relationship between the department and the county. Has the department, in fact, become a contract employee of the county such that the doctrine of *respondeat superior* would apply, making the county liable as an employer? The court also pointed out that because the department had insurance coverage (in this case in excess of $1,000,000), the plaintiff, Mr. Luhmann, would receive payment up to the amount of the insurance coverage, but no more, in accordance with NCGS 153A-435.

In light of the decision in this case, the doctrine of *respondeat superior* could become a critical issue when a department is found to be liable and the department's insurance cannot cover the loss but the combined coverages of the department and the county or municipality with which the department contracted to provide services are sufficient. The wording of the contract as it describes the relationship between the parties could become a very important consideration for the court in a case such as *Luhmann*.

# Corporate (Nonprofit)

In North Carolina, nonprofit corporations are afforded certain defenses to civil liability which are not available to for-profit corporations. This is especially true for volunteer emergency-services providers (VFDs), all of which are non-profit corporations.

The basic rule is that corporate officers and directors are immune from monetary liability for the consequences of their actions except to the extent that insurance coverage exists to cover claims against them. This rule is similar to the insurance exception to sovereign immunity for local government. Corporate officers and directors can be granted immunity by the articles of incorporation (NCGS 55A-2-02(b)(4)) if the incorporators so state in the articles or if the corporation amends its articles accordingly at a later date. In addition, NCGS 55A-8-60 provides immunity for officers and directors. This immunity does not exist if the officer or director:

a. was compensated for services beyond simple expense reimbursement;
b. was acting outside the scope of his authority or duties;
c. was acting in bad faith;
d. committed gross negligence, willful or wanton misconduct, or intentional wrongdoing;
e. derived an improper personal financial benefit from the action in question;
f. incurred the liability as a result of the operation of a motor vehicle;
g. engaged in an unlawful loan or distribution.

These rules concern corporate directors and officers (president, vice president, secretary, treasurer, and so forth) and *not* necessarily the service delivery personnel (fire-rescue or EMS officers and members). Their immunities are covered by other statutes.

Sometimes corporations are accused of acting *ultra vires*—that is, outside the scope of their authority—thereby taking unauthorized actions. Corporations accused of conduct creating liability exposure occasionally seek to "duck the bullet" by arguing that the decision to engage in the activity should not have been made because the corporation had no authority to perform the activity. This argument is known as the defense of *ultra vires*. NCGS 55A-3-04 states that all corporations shall be presumed to have the authority to act in all circumstances and the *ultra vires* assertion places the burden upon the corporation to show that it lacked the authority to act, making it nearly impossible to succeed with such an argument. A director, a member, the Attorney

General, or the legal representative of a receiver or trustee can bring legal action to halt the activity once initiated and can seek monetary damages from the appropriate party. The court can halt the activity by injunction or even set it aside. Using *ultra vires* as a defense is difficult, and many authorities on such matters recommend that it be used only as a last resort. It can be used, however, as a tool (albeit expensive) to control inappropriate decisions by officers or directors.

Beyond these general rules there are a number of additional defenses available to VFDs which are tailored to their activities. The following sections will discuss these.

## General Immunity for VFDs (NCGS 58-82-5)

This statute affords a broad range of civil immunity to a VFD and its members. When engaged in fire suppression or traffic direction or enforcement of traffic laws at the scene of a fire or in connection with a fire, accident or other hazard, the department and its members are immune from liability for the consequences of their actions unless their actions amount to gross negligence, wanton conduct, or intentional wrongdoing. Additionally, any volunteer firefighter or rescue squad member who receives no compensation for his or her services and who delivers EMS to a person at a fire scene is not liable for the consequences of his or her actions unless the conduct amounts to gross negligence, wanton conduct, or intentional wrongdoing.

The Supreme Court of North Carolina has recently issued an opinion clarifying the proper use of the statute. The case began with an accident in the winter involving water spillage at a fill site. The spill resulted in ice, which led to an accident followed by a lawsuit followed by an appeal to the Court of Appeals. The court ruled that because the fill site was one-half mile from the scene of the fire, it was too remote from the scene for the statute to apply. In February 2000 the Supreme Court of North Carolina reversed the findings of the Court of Appeals. The court held that the language of the statute was clear—that as long as the activity in question *related to* the suppression of a reported fire, the statute applied and that, therefore, the department was protected from liability. The relative locations of the activities in question were not important. In reversing the holding of the Court of Appeals, the court pointed out that the clear intent of the statute was to protect volunteer fire departments from the consequences of ordinary negligence when engaged in fire suppression activities (*Spruill v. Lake Phelps Volunteer Fire Department*).

This case is both reassuring and troubling. It is reassuring because the highest court in North Carolina ruled that the statute does precisely what it is supposed to do. The case is troubling because the court made no mention of all the other duties which have been handed to the fire service. The court's language was very careful and limited itself solely to matters concerning responses to reported fires and fire suppression.

To be covered under this statute, a VFD must meet the definition of "rural fire department" according to subsection (a) of the statute. Namely, the VFD must be a nonprofit corporation in North Carolina, must carry an insurance response rating (NCRRS) of not less than 9/9S, and must operate apparatus of a value of not less than $5,000. If the department does not meet these requirements, it fails to fit under the statute's umbrella of immunity. In today's world, incorporation and apparatus requirements are straightforward. However, loss of NCRRS rating is another matter. A poor inspection result leading to loss of a 9S rating, for instance, could have catastrophic consequences for a department. The loss of the rating takes away one of the essential statutory requirements which qualify a VFD for immunity, thereby taking away the immunity itself. The department and its members are thus liable for the consequences of their actions in the same way as ordinary citizens.

Another troubling aspect of this statute is that it concerns only fire and EMS-related activities, while the fire service, both paid and volunteer, is being assigned an increasingly heavy and varied workload. Some lawyers believe that the narrow language of the statute covers any activity ordinarily associated with the fire service; however, the court holdings in the *Geiger* and *Luhmann* cases suggest narrower interpretations of the statute. Given the current fondness of lawyers and courts for hairsplitting, the author believes that it is only a matter of time before a plaintiff attempts to take advantage of these narrower interpretations.

*Geiger v. Guilford College Community Volunteer Fireman's Association, Inc.* (668 F. Supp. 492 [1987]), from the United States District Court for the Middle District of North Carolina, was decided under the predecessor statute to NCGS 58-82-5, NCGS 69-39.1. Both statutes are worded almost identically, with the same intent obviously in mind—liability protection for VFDs. In *Geiger*, the department was summoned to rescue two workers trapped in an above-ground gasoline tank and overcome by the fumes. The rescue was unsuccessful and Mr. Geiger died. The lawsuit was filed in federal court because of jurisdictional issues in the case. The judge ruled that because the department was not engaged in a response to a fire, the statute *did not* protect it. The judge examined the express language of the statute and pointed out (correctly) that the statute dealt only with fire-related activities. In his opinion, Judge Gordon pointed out that it was the responsibility of the General Assembly, and not the courts, to broaden the meaning of so specific a statute. Having said this, however, the judge ruled that a VFD, by contracting with a local government entity to provide fire protection, was providing a governmental service at the express request of a unit of local government and, therefore, was entitled to the protections of sovereign immunity to the extent not covered by insurance, like any other unit of local government.

This case is cited frequently in connection with statutes dealing with various aspects of fire operations. However, it is a ruling in the Middle District of North Carolina only, and the extent of its control over rulings in other courts in North Carolina is not clear. Ultimately, *Geiger* remains an interesting case, but one whose greatest importance may lie in its illustration of the need for legislative action to modify NCGS 58-82-5 to reflect more accurately the current duties of the fire service.

## *Geiger*, *Luhmann*, NCGS 58-82-5, and NCGS 69-25.8

Now VFDs in North Carolina have two cases and two statutes which supposedly describe the departments' civil liability immunities. In *Geiger*, the court held that NCGS 58-82-5 covered activities associated only with responses to reported fires and that VFDs were protected by sovereign immunity in accordance with NCGS 69-25.8. In *Luhmann*, the court held that a department which otherwise met the requirements of the statutory definition of a VFD in NCGS 58-82-5 could have paid personnel but was protected by sovereign immunity in accordance with NCGS 69-25.8. In making this ruling, the court pointed out that it had made a detailed factual study of the department's operations, but the court did not clearly define the facts which led to its ruling. The factors it did discuss were those which characterize a very large percentage of VFDs in North Carolina—a fire-tax district, a large percentage of the budget furnished by fire taxes, contracts with one or more entities to provide services, incorporation as a nonprofit entity to provide the services, and some paid personnel. The only factor which would seem to distinguish Cape Carteret Volunteer Fire and Rescue Department, Inc. from most other VFDs in North Carolina is the number of paid personnel. At the very least, these cases suggest that a VFD which contracts with a municipality or county to provide services may be entitled to the same sovereign immunity defenses as the municipality or county with which it has contracted.

However, the Court, in its opinion, does not distinguish between a fire service district established pursuant to Chapter 153A of the General Statutes and a referendum-based district established pursuant to Chapter 69 of the General Statutes. The language in NCGS 69-25.8 appears to restrict its grant of immunity to referendum-based fire districts. It would appear that the question of which type of immunity protects which type of district remains uncertain unless evidence of how the fire district was created is offered into evidence at trial. An argument could be made that NCGS 58-82-5 protects all VFDs and their members, with the exception of referendum-created districts, which are protected by NCGS 69-25.8. Ultimately, *Luhmann* is a case in which the lawyers for Mr. Luhmann realized that the only way their client could be compensated for his injuries was to convince an appellate court that the proper statute covering his case was NCGS 69-25.8. *Luhmann* certainly "muddies the legal waters" a bit, but it is an "unreported" case, which means the overall ruling is restricted to the facts of the case and is not a broad, overall ruling on a major issue. However, it does raise the importance of discovering and producing as soon as possible in the legal proceedings the method by which the fire district served by the department "at whom the finger has been pointed" was created.

## The Good Samaritan

North Carolina does not require its emergency responders to stop at all emergency scenes they encounter and render assistance, nor does it have a single "Good Samaritan" statute. The state relies primarily on two statutes relating to EMS. One is NCGS 20-166(d), which provides that *anyone* who stops at the scene of a motor vehicle accident and renders aid to an injured person is immune from liability unless the aid rendered amounts to wanton conduct or intentional wrongdoing. The statute makes *no distinctions* regarding the training, employment, or affiliation of the person rendering the assistance.

The second statute is NCGS 90-21.14. Subsection (a) provides immunity for an unpaid volunteer EMS worker who renders aid to a person who is ill, unconscious, or injured when it is apparent that failure to render the aid either will cause or is likely to cause deterioration in the patient's condition. The statute specifically does not apply to someone rendering EMS care in the course of his or her business or profession. If NCGS 90-21.14 is found to be in conflict with NCGS 20-166(d), the provisions of NCGS 20-166(d) apply, making assistance at a motor vehicle accident a situation where the rules are very generous.

A statutory cousin of the foregoing two statutes is NCGS 58-82-5(c). This one affords protection to volunteer firefighters or volunteer EMS workers who deliver emergency care to unconscious, ill, or injured persons at the scene of a fire. They are protected unless their behavior is found to constitute gross negligence, wanton conduct, or intentional wrongdoing. "Volunteers" are defined as those delivering the EMS who do not receive compensation for their services.

## Emergency Management (NCGS 166A-14)

The statute states that all emergency management activities are deemed governmental functions and that North Carolina and any political subdivision thereof shall be immune from liability for damages resulting from the death or injury of any person or injury to any property which may result from actions undertaken in accordance with any emergency management statute, rule, regulation, or ordinance. A worker (or the agency) can be held accountable only if the action is found to have been taken in bad faith or under circumstances involving gross negligence or intentional wrongdoing. The statute applies to paid, volunteer, or auxiliary personnel who perform services in accordance with any agreement for emergency management assistance anywhere in North Carolina.

## Volunteers (NCGS 1-539.10)

This statute established blanket immunity for volunteers who work for charitable organizations. As long as the work is undertaken in good faith and under reasonable circumstances, and does not involve gross negligence, wanton conduct, intentional wrongdoing, or the operation of a motor vehicle, the volunteer is immune. However, if the organization for which the volunteer works has liability insurance, the immunity is waived to the extent that insurance coverage exists. According to NCGS 1-539.11, a volunteer is a worker who receives no compensation for his or her services other than expense reimbursement, and a charitable organization is a nonprofit organization which has humanitarian and philanthropic objectives and which conducts activities that benefit humanity or a significant segment of the community. While this statute does not appear to have been used in a case involving a VFD, it remains available as a possible defense tool.

A federal equivalent is the Volunteer Protection Act (VPA) of 1997 (42 U.S.C.A. 14501–14505). It is designed to protect from liability for the consequences of ordinary negligence people who volunteer their services to local government or to nonprofit entities which are properly tax-exempt under the Internal Revenue Code. It does not protect them for the consequences of

1. criminal conduct;
2. intentional acts;
3. reckless conduct;
4. gross negligence;
5. acts not within the scope of the volunteer's duties;
6. the operation of a motor vehicle for which the state requires a license or insurance;
7. activity for which the state requires a license or certification which the volunteer lacks;
8. an injury occurring while the volunteer is under the influence of drugs or alcohol;
9. violent crimes;
10. hate crimes;
11. a sexual offense for which the volunteer has been convicted in any court;
12. activities which violate any state or federal civil rights law.

The law does not appear to have been applied in a fire service context in North Carolina as yet. Whether or not it would be applied is problematic because it is a federal statute and the contexts within which it might be applicable are unclear.

## Common Law Defenses (Court-Established Rules)

These are defenses available in tort liability cases which can ordinarily be applied regardless of the identity of the defendant. They are the result of court rulings and/or the application of English common-law rules which became part of North Carolina law. These rules can possess many complex interpretations because they are used frequently in court in many different situations. This discussion will cover the basics.

### *Contributory Negligence*

The theory behind contributory negligence is that the law imposes a duty upon all of us to use due care to protect ourselves (care consistent with the risk involved) and to take reasonable steps to avoid injury when we are aware of a danger or should reasonably be aware of a danger. A person who is not an adult, or who is mentally incompetent, is held to a different standard of self-protection, depending upon the circumstances of the case. North Carolina's courts have stated quite clearly that a person cannot recover damages for an injury he or she helped cause.

Contributory negligence occurs when an act or omission of the plaintiff becomes all or part of the proximate cause of the plaintiff's injury by the defendant. In other words, something the plaintiff did helped to cause or did, in fact, cause his or her injury. The contributory activity by the plaintiff does not have to be the sole cause; it need only play a part in the activities which led to the injury. Because North Carolina does not compare relative levels of negligence between plaintiff and defendant when determining liability, the plaintiff has to contribute only a minuscule amount to his injury to lose. The rule is an absolute one. Carelessness can be contributory negligence. North Carolina's courts have held that a plaintiff who ignores a risk or danger of which he or she is aware, or should reasonably be aware, has participated in the injury to some degree and committed contributory negligence. The case *Willis v. Town of Beaufort* was resolved by contributory negligence. Mr. Willis, a commercial fisherman, went belowdecks on his wooden fishing boat and attempted to weld a fuel tank. The boat caught fire and burned to the waterline. Mr. Willis sued the Town of Beaufort and the case was dismissed during the trial. After being appealed, the case came back to Carteret County and went to a jury trial. In its verdict, the jury found that the plaintiff was contributorily negligent, and no money was awarded to the plaintiff.

One circumstance recognized by the courts as a situation when contributory negligence, even when present, will not be applied is willful and wanton (extreme) conduct by a defendant. Certain kinds of criminal acts or intentional disregard for others are examples of this type of behavior. However, as with all situations of contributory negligence, each case is reviewed by a court on an individual basis.

Another doctrine relating to contributory negligence is called "Last Clear Chance." This doctrine says that even though the plaintiff was negligent and the defendant was contributorily negligent, the conduct of the defendant will not be held against him because the plaintiff became aware of the problem, had a last reasonable opportunity to take steps to avoid or prevent the injury, but went ahead anyway and caused the injury. Additionally, the defendant must have had the time and means to avoid the injury. The opportunity to avoid it must be a clear one, not merely a possibility. The *Luhmann* case was decided by the jury on a matter of last clear chance. The evidence showed that the driver had an opportunity to walk around the tanker and determine the locations of personnel and equipment prior to departing the fill site and thereby avoid the danger posed by the proximity of the plaintiff to the equipment but that he failed to do so. Other evidence suggested that no one from the department asked the plaintiff to move out of the area of danger prior to his injury.

The Doctrine of Rescue says that if someone is injured while attempting a bona fide rescue of another from imminent peril there can be no contributory negligence on the part of the rescuer if the rescue was attempted and was not performed rashly or recklessly. Additionally, the rescuer must not have been part of the cause of the original endangerment.

### *Assumption of Risk*

This a doctrine is often confused with contributory negligence. In a sense, the plaintiff who goes ahead even though reasonably aware of the danger has "assumed" the risk of the consequences of his or her actions. However, North Carolina makes an important distinction between assumption of risk and contributory negligence. One cannot assume a risk within the meaning of the doctrine unless there is a contractual relationship between the plaintiff and defendant. A structure fire could be an example of this. The responding firefighters have contractual relationships with their departments and their workmen's compensation insurance carriers who have agreed to protect them from injury or compensate them if they are injured. By initiating an interior attack, the firefighters have accepted, or assumed, the risks associated with such an activity and cannot take legal action against their department in case of death or injury unless it can be shown that the department caused an extreme set of circumstances which led to the injury or death (a *Woodson*-type set of circumstances.)

### *Sudden Peril or Emergency*

This doctrine says that a person who is suddenly and unexpectedly confronted with an emergency which he or she did not cause is not expected to use the same judgment when coping with the problem as would be expected of someone with the opportunity and time to observe what is happening and plan a response. The court will expect the defendant to use only the degree of care which could be expected of a reasonable and prudent person under the same or similar circumstances. It does not matter that in hindsight a better decision could have been made. As long as the defendant is found to have acted as a reasonable and prudent person at the time of the event, this kind of defense can take effect. Fire ground situations are typically little other than series of sudden emergent events, requiring quick decisions under stressful circumstances. A fire department is expected to cope with these occurrences more ably than the average citizen because of training and equipment. However, the defense of sudden peril or emergency should remain available because its use by the courts is based upon the facts and circumstances of each case. A fire ground evolution sometimes gets away from those running the scene even though everything was done correctly from the beginning. Hindsight notwithstanding, it would seem fair to assume that a response to the event, if reasonable under the circumstances at the time, would allow the application of the doctrine.

### *Acts of God*

An "Act of God" is an event which occurs only as a result of an act of nature, without any human influence. It must be a natural occurrence, and if the cause of the injury is an act of God, the plaintiff cannot recover any damages. However, when the act of God is combined with someone else's negligence, the plaintiff may recover damages. This means that an act triggered by nature can become the basis of a lawsuit if someone's negligence becomes part of the proximate cause of the injury.

### *Age*

North Carolina, like most states, takes age into consideration when evaluating the conduct of a person, especially when liability is a consideration. The courts presume as a matter of law that a child under the age of seven years cannot commit contributory negligence and that one aged seven to fourteen carries a rebuttable presumption that he or she cannot commit contributory negligence. This means that the court begins by assuming, as a matter of law, that the child is incapable of committing contributory negligence but the defendant has the option of trying to prove that the child is capable of committing contributory negligence and that he or she did, in fact, do so.

A person over the age of eighteen is legally an adult in North Carolina and therefore liable for the consequences of his or her actions unless the court can be satisfied that the person was under a sufficient disability such that the person was not in control of himself or herself at the time of the event. This disability cannot be one which the defendant brought on himself or herself (for example, DWI).

Parents cannot waive potential liability claims for their minor children in North Carolina. The waiver can be effective for any parent who signs it, but the minor child has until one year after his or her eighteenth birthday to file a lawsuit regarding any claims for personal injuries sustained while a minor. There is an old saying among lawyers in North Carolina that a waiver of liability has never been written which could not be demolished by a competent trial lawyer. This is something all departments which operate junior programs should keep in mind.

Ordinarily, injury to or death of a child in which one or both parents' contributory negligence is established prevents the parents from recovering damages as a result of the injury or death. However, if it can be shown that only one parent was contributorily negligent, courts have allowed the other parent to recover damages.

## The Role of Insurance

As has been pointed out already, the presence of insurance opens the door to departmental liability at trial, but the exposure, in most instances, is limited to the amount of the insurance coverage. Today, good insurance coverage remains the best protection available to departments and their members. It is impossible to perform all evolutions associated with service delivery perfectly every time, regardless of training or experience levels within a department. With the increasingly complex evolutions required even of the smaller rural VFDs, the likelihood of a mistake causing an injury will only increase with time. The longer a department exists, the greater the likelihood there is of a mistake. It's a fact of life.

Insurance accomplishes two things. First, it compensates a member of the public who has been injured, which is just plain good public relations. It also pays legal expenses. North Carolina insurance law requires that all companies doing business in the state defend all claims against their insureds. The cost of defending a lawsuit can be very high, easily reaching into the tens of thousands of dollars, and lawyers who defend lawsuits are paid as they go along, unlike lawyers who file them, who, ordinarily, take a case on a contingent basis (they roll the dice and are paid only if they win the case). A poorly insured department can spend great amounts of its or the taxpayers' money defending a lawsuit, with resulting taxpayer unhappiness at budget time.

# From a Lawyer's Viewpoint

The following remarks are based upon the writer's experience in the fire service and as a practicing attorney and are not intended to be all-inclusive or legally binding upon anyone. However, it is hoped that they will add some small amount of perspective to the foregoing discussion of liability. The prospect of a lawsuit or claim for damages is now a fact of life with which we all must learn to live. Today, if someone gets mad enough, he or she will try to sue. This situation is exacerbated by the general misapprehension that all problems can be solved in court. While this certainly is not the case, the learning of that lesson can be an expensive and unpleasant experience for all concerned.

Unfortunately, this fondness for litigation has not been discouraged adequately in this state and may never be so. To some extent, this is because every year more persons are qualified as lawyers and are sent out into the world to earn a living and there is available to them only a limited amount of economic "pie" which they must divide. As a result, lawyers develop novel theories to establish liability and take cases which sometimes they wish they had refused (these cases are known as "dogs"). Once a lawyer takes a case, North Carolina does not allow him or her to get out of it except by carrying the case to its conclusion, obtaining the consent of the client (which seldom happens because the client is convinced the "dog" of a case is the solution to all his economic woes and will permit him to retire and live in luxury for the rest of his days), or obtaining an order from the court permitting withdrawal (which ordinarily is granted only sparingly).

Practicing law is not an inexpensive business. The overhead for a law office, especially one engaged in trials, is very high. Lawsuits require a lot of paperwork as well as competent (expensive) support staff. The only way to pay the bills, and hopefully themselves, is for lawyers to take cases and work them vigorously. The North Carolina State Bar actively enforces a lengthy set of rules which govern the conduct of lawyers. Among these rules is an absolute requirement that a lawyer represent a client's interests vigorously. The idea is to prevent lawyers from simply taking cases and playing with them (another stimulus for legal activity) and to assure as much as possible that the client is paying for good legal services.

Periodically, studies relating to the practice of law are conducted and some of the results can be amusing. One recent study of clients found that at least 90 percent of them believed that their lawyers were representing them well and were good people. However, approximately the same percentage believed that the lawyers on the other side were lower forms of life and not nice folks.

So what do you do if there is a problem? First, shut up! Next, notify the command structure and your insurance company. Then follow their instructions. The insurance company will offer guidance in damage control, public relations, and investigation of the circumstances. If criminal activity is suspected, notify law enforcement after consultations with an attorney. Continue to shut up! This is difficult, but it helps with both damage control and investigation of the facts. Talk creates witnesses and rumors, both of which can be harmful to your case. If you must talk with someone, the law allows certain conversations to be kept private, or privileged. These include conversations with a spouse, one's attorney, physician, psychologist, or a member of the clergy if the purpose of the conversation is counseling regarding the events in question.

If witnesses are involved, make a list of all witnesses of which you may be aware, furnish it to your attorney, and update it as needed. Finally, if you as an individual believe you are being targeted, consult an attorney of your own.

One other thing to remember is that most lawyers are not familiar with fire departments, how they function, and some of the relationships between the members of the fire service which often result from service in departments. Be patient; let the lawyer explain his or her view of the circumstances and then explain them from your perspective. Hopefully, your lawyer will be a good listener and take your views into account.

# Chapter 17

# Social Media

Because social media are widespread and easily available to computer-literate members of emergency services, the possibility of abuse, or at the very least, inappropriate, use of social media by emergency services personnel has become an important concern within the emergency services community. There exist few legal guidelines for employers or supervisors to manage social media, leaving the fire service with few options other than to develop our own guidelines until legislative bodies or the courts furnish us with guidance.

Using social media is a form of speech or expression which is protected from governmental interference by the freedom of speech and expression clauses of the First Amendment and Article I, Sections 12, 13, and 14 of North Carolina's constitution. This general category of expression is called "protected speech" by the courts, meaning that it is protected from any form of governmental interference. However, over time, our courts have established some exceptions to the general rule of protected speech, thereby allowing government to regulate them.

Unauthorized disclosure of classified governmental information and dissemination of child pornography are not protected speech. Sometimes disclosure of industrial trade secrets is not regarded as protected speech. Harmful speech is sometimes regarded as "unprotected" and subject to governmental regulation and penalties. This is speech or expression created with the intent of causing harm to persons or property or speech so irresponsible as to be obviously harmful (yelling "Fire!" in a crowded movie theater just so see what will happen).

Because the fire service routinely encounters situations (fires, rescues, motor vehicle accidents) which the average citizen may never encounter in a lifetime, the availability of social media when combined with the inherently unusual nature of fire service operations creates an environment which some members of the fire service (especially younger ones) find too tempting to ignore. The usual pattern of behavior is to create images of these events and post them on social media.

Sometimes, these images depict activities of the responders which are inappropriate, which may violate existing standards of performance or decision-making, or which may violate the law. Other times they may depict very private moments in the victims' lives or in those of their families. If the event in question involves medical matters, the posting is a colossal violation of HIPPA. The posted material becomes evidence admissible against the department and its members in civil litigation, regulatory hearings (e.g., OSHA, HIPPA), and possible criminal prosecution. Every responding department has a duty to perform its duties competently, within whatever regulatory boundaries which may apply to the evolution, while affording due respect to the privacy of those being served. When those duties are violated and someone or someone's property is injured or a HIPPA rule is violated, the department, and sometimes one or more of its members, may be held accountable in court.

So what has happened when the inappropriate posting to social media is performed? Obviously the proverbial "manure storm" has erupted, but that may be only the beginning. Not only has the department been caught "red-handed" violating a regulation or law, it has violated the privacy of a patient or victim and embarrassed or humiliated a family. This behavior damages a department's relationship with its constituents and the governmental agency of which it is a part or with which it has a contract. The potential consequences are not difficult to imagine.

Another destructive social media activity is members of a department posting uncomplimentary or untrue messages relating to other members, the department itself, other departments, or town or county commissioners. While these postings may appear to be entertaining to the person posting them, they can be destructive to the department's operational capabilities and the department's relationships with other departments, its constituents, and the commissioners upon whom the department depends for its support. This undermines departmental leadership and the department's relationship with the governmental agency with which the department has a contract. The author has witnessed this process on more than one occasion.

The responsibility for preventing social media catastrophes lies with the leadership of each department, which has the difficult task of balancing constitutionally-protected freedom of expression rules against the need to protect the privacy of its constituents. The fire service is one of the branches of emergency services which has been deemed to be a custodian of something called the "public trust." This means that because they are authorized to take emergency actions to protect the public, the law affords them special privileges to perform their duties, but they, in turn, are expected to respect the public and protect its privacy. This position of public trust also allows those agencies holding the public trust to impose regulations on themselves in order to carry out this obligation. However, these rules must be managed in a fashion which does not violate the constitutional rights of those performing the services. Unfortunately, the line between constitutionality and unconstitutionality is not always clear in this context, so experimentation and possibly litigation to "finally" answer the question can be the result. However, "ya' gotta start somewhere," as the saying goes.

Bylaws, SOG's, personnel policies, and employment contracts should include rules relating to the use of social media while on the job or relating to on the job activities, accompanied by meaningful penalties for violations. For example, forbidding the carrying of imaging equipment by members (except as authorized by command authority) to event scenes, prohibiting the use of departmental signs, badges, seals, or other identifying information on any postings on the internet (unless it is official business, obviously), prohibiting the posting of information intended to bring discredit upon or embarrassment to the department (see Article 134, UCMJ) could be included in departmental governing or operational documents, to be followed by written acknowledgment by each members that he or she has read, understands, and agrees to abide by the policy. The existence of these rules accompanied by vigorous enforcement should help restore the department's image and working relationships should a social media disaster erupt.

Arguments currently are being made and questions asked regarding how restrictions social media use relate to the First Amendment. Remember, the First Amendment restricts the ability of government to interfere with a person's rights of self-expression. An employer can impose those restrictions as a condition of employment and subject a violator of those rules to disciplinary action. However, the employer should document those rules as described in the previous paragraph. The same ideas and rules could be implemented in a VFD, as well, again the department's leadership should make certain that all departmental members (and employees, if any) are informed of those rules.

The late Tom Clancy wrote a number of meticulously researched novels and books dealing with defense and national security affairs. His successors, working under his name have continued to produce novels dealing with national security matters and not long ago published a novel entitled *True Faith and Allegiance*. One of the leading villains of the story eventually is identified as a computer hacker from central Europe, who is apprehended and interrogated by American authorities. In the course of his interrogation, he describes how he uses social media to learn virtually anything he wants to learn regarding the personal lives and finances of his "targets," subsequently furnishing this information his "employers" for them to use against the best interests of the United States. The disturbing aspect of what he reveals is the ease with which he explores people's lives by utilizing their social media accounts. This means, in theory at least, that any member of emergency services who participates in social media while carrying sensitive or private information on his or her devices or computers could easily be hacked and that information copied and utilized for unintended purposes by people other than the owner of the hacked social media account. This is a serious security risk when one considers the role played by the fire service in responses to attacks on the Homeland.

In October 2018, the author attended the annual meeting of the North Carolina and South Carolina chapters of the International Association of Arson Investigators. One of the presenters during the five-day training session was an engineer and former member of the United States Secret Service who presently is a Special Agent and investigator for the office of the Inspector General of the United states Postal Service. His topic was cellphone forensics and during the presentation we learned that when one plugs a device into the USB port in a motor vehicle, the entire contents of the device are downloaded into the vehicle's computer system, thereby demolishing the privacy and security of that information.

The author believes that the unique nature of emergency services enables a strong argument to be made to a court that the privacy rights of those protected by restrictions on freedom of expression within the department outweigh the rights of those who might desire to post the information in question. Congress made this idea abundantly clear in the privacy protection provisions of HIPPA. Now the fire service must determine how to apply privacy rules in a slightly different context. After all, once it's "out there," it can't be "called back."

# Bibliography

Allred, Stephen. *Employment Law*. Chapel Hill, NC: Institute of Government, 1992.

Avram, Randall, Albert R. Bell, Jr., Maria Hallas, M. Robin Davis, and Jonathan W. Yarborough. *Basic Wage and Hour Law in North Carolina*. Eau Claire, WI: National Business Institute, 1998.

Blackstone, Steve. "HIPPA & the Fire-EMS Service: The Early Returns." *Firehouse*. August 2003.

Brown-Graham, Anita R. *A Practical Guide to the Liability of North Carolina Cities and Counties*. Chapel Hill, NC: Institute of Government, 1999.

Gray, S. McKinley, III, Randall D. Avram, Todd W. Cline, and Martha P. Brown. *Fundamental Issues in North Carolina Human Resources Law*. Eau Claire, WI: National Business Institute, 1999.

Lawrence, David M. *Local Government Finance in North Carolina*. Chapel Hill, NC: Institute of Government, 1990.

__________. *Open Meetings and Local Governments in North Carolina*. Chapel Hill, NC: Institute of Government, 2002.

Loeb, Ben F., Jr. *Fire Protection Law in North Carolina*. 5th ed. Chapel Hill, NC: Institute of Government, 1993.

Ludwig, Gary. "Beauty and the Beast, More About HIPPA." *Firehouse*. August 2003.

*North Carolina Fire and Emergency Services Law, Annotated*. Chapel Hill, NC: Institute of Government, 2003.

*Strong's North Carolina Index*. 4th ed. N.p.: West Group, 1999.

*West's North Carolina Digest*. 2nd ed. St. Paul, MN: West Publishing, 1996.

Wilbur, Michael. "Updates: The Year in Review." *Firehouse*. November 2003.

# Index